SCHEMES

The Language of
Modern Algebraic Geometry

The Wadsworth & Brooks/Cole Mathematics Series

Series Editors
Raoul H. Bott, Harvard University
David Eisenbud, Brandeis University
Hugh L. Montgomery, University of Michigan
Paul J. Sally, University of Chicago
Barry Simon, California Institute of Technology
Richard P. Stanley, Massachusetts Institute of Technology

M. Adams, V. Guillemin, *Measure Theory and Probability*
W. Beckner, A. Calderón, R. Fefferman, P. Jones, *Conference on Harmonic Analysis in Honor of Antoni Zygmund*
G. Chartrand, L. Lesniak, *Graphs & Digraphs, Second Edition*
W. Derrick, *Complex Analysis and Applications, Second Edition*
J. Dieudonné, *History of Algebraic Geometry*
R. Dudley, *Real Analysis and Probability*
R. Durrett, *Brownian Motion and Martingales in Analysis*
D. Eisenbud, J. Harris, *Schemes: The Language of Modern Algebraic Geometry*
R. Epstein, W. Carnielli, *Computability: Computable Functions, Logic, and the Foundations of Mathematics*
S. Fisher, *Complex Variables, Second Edition*
G. Folland, *Fourier Analysis and Its Applications*
A. Garsia, *Topics in Almost Everywhere Convergence*
P. Garrett, *Holomorphic Hilbert Modular Forms*
R. Gunning, *Introduction to Holomorphic Functions of Several Variables*
    Volume I: Function Theory
    Volume II: Local Theory
    Volume III: Homological Theory
H. Helson, *Harmonic Analysis*
J. Kevorkian, *Partial Differential Equations: Analytical Solution Techniques*
S. Krantz, *Function Theory of Several Complex Variables, Second Edition*
R. McKenzie, G. McNulty, W. Taylor, *Algebras, Lattices, Varieties, Volume I*
E. Mendelson, *Introduction to Mathematical Logic, Third Edition*
D. Passman, *A Course in Ring Theory*
B. Sagan, *The Symmetric Group: Representations, Combinatorial Algorithms, and Symmetric Functions*
R. Salem, *Algebraic Numbers and Fourier Analysis* and L. Carleson, *Selected Problems on Exceptional Sets*
R. Stanley, *Enumerative Combinatorics, Volume I*
J. Strikwerda, *Finite Difference Schemes and Partial Differential Equations*
K. Stromberg, *An Introduction to Classical Real Analysis*
W. Taylor, *The Geometry of Computer Graphics*

SCHEMES
The Language of Modern Algebraic Geometry

David Eisenbud

Brandeis University

Joe Harris

Harvard University

Wadsworth & Brooks/Cole Advanced Books & Software

Pacific Grove, California

QA
E36
1992

DISCARDED

LIBRARY
Appalachian State University
Boone, North Carolina 28608

DISCARDED

Wadsworth & Brooks/Cole Advanced Books & Software
A Division of Wadsworth, Inc.

© 1992 by Wadsworth, Inc., Belmont, California 94002.
All rights reserved.
No part of this book may be reproduced, stored in a retrieval system, or transcribed,
in any form or by any means—electronic, mechanical, photocopying, recording,
or otherwise—without the prior written permission of the publisher,
Brooks/Cole Publishing Company, Pacific Grove, California 93950,
a division of Wadsworth, Inc.

Printed in the United States of America

10 9 8 7 6 5 4 3 2 1

Library of Congress Cataloging in Publication Data
Eisenbud, David.
    Schemes : the language of modern Algebraic Geometry / David
Eisenbud and Joe Harris.
        p. cm.
    Includes bibliographical references and index.
    ISBN 0-534-17606-2.—ISBN 0-534-17604-6 (pbk.)
    1. Group schemes (Mathematics) I. Harris, Joe, II. Title.
QA564.E36 1992
516.3′5—dc20   91-45534
                                                        CIP

Sponsoring Editor   *John Kimmel*
Marketing Representative   *John Moroney*
Editorial Associate   *Nancy Miaoulis*
Production Editors   *Kay Mikel and Nancy Shammas*
Manuscript Editor   *Carol I. Beal*
Permissions Editor   *Carline Haga*
Interior and Cover Design   *Roy R. Neuhaus*
Art Coordinator   *Lisa Torri*
Interior Illustration   *Gloria Langer*
Typesetting   *Asco Trade Typesetting Limited*
Cover Printing   *Phoenix Color Corporation*
Printing and Binding   *Arcata Graphics/Fairfield*

He had bought a large map representing the sea,
    Without the least vestige of land:
And the crew were much pleased when they found it to be
    A map they could all understand.

From "The Hunting of the Snark" by Lewis Carroll

Preface

This text is intended to fill the gap between texts on classical algebraic geometry and the full-blown accounts of the theory of schemes.

Currently, the student with a background in the classical theory must wade through a vast technical development before understanding the reasons and elementary examples that underlie the current language in which algebraic geometry is done: the theory of schemes. This barrier may prevent the mathematician with a casual interest in algebraic geometry from understanding many modern developments.

We have tried to help the reader over this barrier by focusing on the definitions and some of the elementary examples that lead to the ideas of scheme theory. For example, we spend time on the deep analogy between curves and rings of integers in number fields, which scheme theory treats, finally, as two instances of the same kind of object; and we also treat at some length the notion of flat families, and their limits, which motivates very persuasively the introduction of nilpotent elements.

The prerequisites for reading this book are modest: a little commutative algebra (localization, prime ideals, the Noetherian condition) and an acquaintance with algebraic varieties, roughly at the level of a semester course. Among the texts providing more than enough commutative algebra, we mention Atiyah-Macdonald (1969), Kunz (1985), and Eisenbud (1992); suitable sources for the algebraic geometry, again providing much more than we actually use, would be Reid (1990), chapters one and two of Shafarevich (1974), chapter one of Hartshorne (1977), the book of Mumford (1976), and Harris (1992).

Many people have helped us with comments on and criticisms of this manuscript, and we benefited from the stalwart efforts of students at both Brandeis and Harvard to understand early versions. Joe Buhler, Herbert Clemens, Vesselin Gasharov, Benedict Gross, Kurt Mederer, and Irena Peeva have contributed particularly helpful lists of comments and complaints.

Finally, Keith Pardue did a wonderful job of proofreading; naturally, the errors that remain are ours.

Finally, we would like to credit David Mumford as the source of inspiration for this book. At one time we planned to write a descendant of the "Red Book," David Mumford's *Introduction to Algebraic Geometry* (preliminary version of the first three chapters) (1960). This would have been based on notes of Mumford (partially revised by Serge Lang) that were originally intended as the second volume of Mumford's *Algebraic Geometry I* (1976). However, as we worked on this project, we began to feel that a systematic textbook of the sort originally envisioned was not really what was needed anymore. The book of Hartshorne (1977), for example, gives an extremely smooth and polished account of a good part of the technical material we had planned to include.

We thus owe an enormous debt to Mumford. Many of the ideas, pictures, and tricks of presentation here come from him, as anyone who attended his lectures on the subject will recognize. For all this, and for his help and mathematical support, we are deeply grateful.

David Eisenbud
Joe Harris

Contents

III   Projective Schemes   84

IV   The Functor of Points   119

SCHEMES
The Language of Modern Algebraic Geometry

Introduction

The theory of schemes is the foundation for algebraic geometry proposed and elaborated by Alexander Grothendieck and further developed by a large circle of other mathematicians who worked with and around him. Initially described in the 1950s and amplified over the course of the succeeding two decades, the theory now permeates algebraic geometry. Since its development as a foundation, the theory has proved extremely successful and flexible. It has led to a true integration of algebraic number theory and algebraic geometry, fulfilling the dreams of earlier generations of number theorists. Many classical problems have been solved. In developing the theory, Grothendieck was partly aiming at a major series of conjectures of André Weil, and indeed this program was finally successful: the proof of the "Weil Conjectures" was completed by Pierre Deligne (1974) using scheme-theoretic tools. But the theory has extended far beyond its initial motivations. Perhaps its most luminous recent success is in the proof of the Mordell Conjecture by Faltings (see, for example, the book of Cornell and Silverman [1986]).

Though the deepest uses of scheme theory at the moment are surely in the number-theoretic direction, elementary "scheme theoretic thinking" is fundamental to a great deal of the current work on classical algebraic geometry. The precise and flexible ideas of degenerations and families to which it has given rise have been central in the solutions of many outstanding problems in the theory of curves, a modern version of the theory of surfaces, and the remarkable new work on higher-dimensional varieties recently undertaken by Mori and others. Some of this work could have been formulated without schemes (though other parts, such as Mori's celebrated proofs involving reduction mod p, use the techniques of scheme theory in an essential way), but all of it is very much informed by the language and concepts of scheme theory.

Despite all this, the average mathematician, and even many a beginner in algebraic geometry, still considers the theory of schemes as something horribly technical and probably irrelevant to what he or she is doing. This is partly because the theory is so general—anything that unifies all of number theory and geometry has got to have a lot of special hypotheses to deal with!—and also partly because of the nature of the textbooks currently available. The pioneering work of Grothendieck and Dieudonné, the *Éléments de Géometrie Algébrique* (1960–1967), became so encyclopedic that it was abandoned (but not before several supplements to Chapter 0 had to be written to accommodate a rapacious need for generality). The book of Hartshorne (1977) provides a highly polished source for the basic theory, and the books of Shafarevich (1974), Iitaka (1982), and others all have something to recommend them; but there still has been a need for a text that would focus on interesting examples, with a minimum of machinery, to show "what was going on" in the field.

This is the book we have tried to write. We felt it of foremost importance that the book remain short, since the beginner needs to move forward, after not too many weeks of playing around without technique, and learn the powerful theories that have been built to handle schemes. This limitation of space (and also our own limitations in expertise and energy) has prevented us from adding many desirable topics: intersection theory (perhaps including a taste of the new Arakelov theory) (Fulton [1984], Cornell and Silverman [1986]), zeta functions (Thomas [1977]), toric varieties (Kempf et al. [1973], Danilov [1978], Oda [1985]), cohomology theories of various sorts (Milne [1980]), and many other topics tempted us but were abandoned. We hope that what we have included is nevertheless useful.

The detailed contents of this booklet are as follows:

Chapter I lays out the definition of schemes, their morphisms, and a few of the most elementary relations, culminating with a discussion of fibered products. We have tried to explain the definitions and their motivations fully, while keeping the section as brief as possible. Thus we have included only those things necessary for the following chapter, which consists entirely of examples.

Chapter II takes up several families of examples of affine schemes and forms the heart of this book. We have focused on affine schemes because virtually all of the differences between the theory of schemes and the theory of abstract varieties are encountered in the affine case—the general theory is really just the union (*pushout* is perhaps more appropriate) of the theory of abstract varieties à la Serre and the theory of affine schemes. We begin with the schemes that come from varieties over an algebraically closed field (II-A). Then we drop various hypotheses in turn and look successively at cases where

the ground field is not algebraically closed (II-B),
the scheme is not reduced (II-C),
the scheme is not defined over a field at all (II-D).

One of the major ways in which nonreduced schemes occur is as limits in flat families. In Section II-C we define various notions related to flatness and explain how to take flat limits of families of subschemes.

Projective schemes form the most important family of nonaffine schemes, indeed the most important family of schemes altogether, and Chapter III is devoted to these. After a discussion of *properness*, which is one of the essential features of projective schemes, we give the construction of Proj and describe in some detail the examples corresponding to projective space over the integers and to the various possible projective double lines in 3-space (in affine space all double lines are equivalent, as we show in Chapter II, but this is not so in projective space).

The remainder of Chapter III is devoted to some simple invariants of projective schemes. We define free resolutions, graded betti numbers, and Hilbert functions, and we study a number of examples to see what these invariants bring in simple cases. We also return briefly to flatness and describe its relation to the Hilbert polynomial.

In Chapter IV we take up a subject peculiar to the theory of schemes, the "functor of points," and we give some of its varied applications: to finding the properties of a scheme, to computing its tangent spaces, to proving existence, to defining extra structure on it, and finally to solving moduli problems. We also give a taste of the way in which some apparently geometric definitions such as that of a tangent space or of openness can be extended from the context of schemes to the context of certain functors. This extension represents the beginning of the program of enlarging the category of schemes to a more flexible one, which is akin to the idea of adding distributions to the ordinary theory of functions.

Since we believe in "learning by doing," we have included a large number of exercises, spread throughout the text. Their level of difficulty and the background they assume varies considerably, so we have included hints to many. The exercises for which there are hints are indicated with a * in the text, and the hints will be found in the Appendix to the book. We feel that the function of this book will be best served if it is read fairly quickly, so we advise the reader to try all the exercises but to turn fairly soon to the hint on those that do not seem straightforward.

Basic Definitions

Just as topological or differentiable manifolds are made by gluing together open balls in Euclidean space, schemes are made by gluing together open sets of a simple kind, called *affine schemes*. There is one major difference: in a manifold one point looks locally just like another, and open balls are the only open sets necessary for the construction; they are all the same and very simple. By contrast, schemes admit much more local variation; the smallest open sets in a scheme are so large that a lot of interesting and nontrivial geometry happens within each one. Indeed, in many schemes no two points have isomorphic open neighborhoods. We will thus spend a large portion of our time describing the affine schemes.

We will lay out basic definitions in this chapter. We have provided in this chapter a series of easy exercises embodying and applying the definitions. The examples given here are mostly of the "simplest possible example of the phenomenon" sort and are not necessarily typical of interesting geometric examples. The next chapter will be devoted to examples of a more representative sort. These are intended to indicate the ways in which the notion of a scheme differs from that of a variety and to give a sense of the unifying power of the scheme-theoretic point of view.

A Affine Schemes

An *affine scheme* is an object made from a commutative ring. The relationship is modeled on, and generalizes, the relationship between an affine variety and its coordinate ring. In fact, one can be led to the definition of scheme in the following way. The basic correspondence of classical algebraic geometry is the bijection

$$\{\text{affine varieties}\} \longleftrightarrow \left\{\begin{array}{c}\text{finitely generated, nilpotent-}\\ \text{free rings over a field } k\end{array}\right\}$$

Here the left-hand side corresponds to the geometric objects we are naively interested in studying—that is, the zero loci of polynomials. If we start by saying that these are the objects of interest, then we arrive at the restricted category of rings on the right. Scheme theory arises if we adopt the opposite point of view: if we do not accept the restrictions "finitely generated," "nilpotent-free," or "k-algebra" and insist that the right-hand side include all commutative rings, what sort of geometric object should we put on the left? The answer is "affine schemes"; and in this section we will show how to extend the preceding correspondence to a diagram

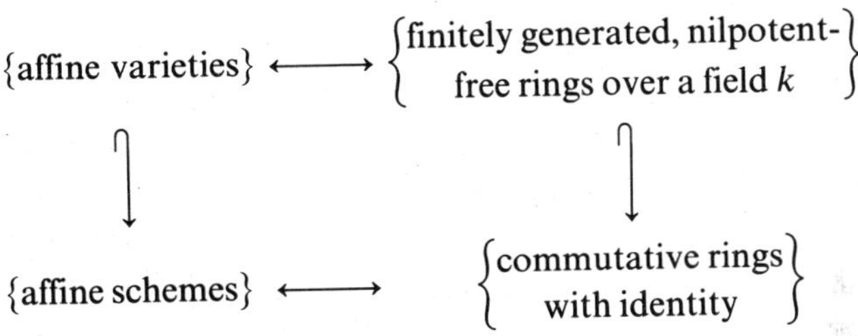

We shall see that in fact the ring and the corresponding affine scheme are equivalent objects. The scheme is, however, a more natural setting for many geometric arguments; speaking in terms of schemes will also allow us to globalize our constructions in succeeding sections.

Looking ahead, the case of differentiable manifolds provides a paradigm for our approach to the definition of schemes. A differentiable manifold M was originally defined to be something obtained by gluing together open balls—that is, a topological space with an atlas of coordinate charts. However, specifying the manifold structure on M is equivalent to specifying which of the continuous functions on any open subset of M are differentiable. The property of differentiability is defined locally, so the differentiable functions form a subsheaf $\mathscr{C}^\infty(M)$ of the sheaf $\mathscr{C}(M)$ of continuous functions on M (the definition of sheaves is given below.) Thus we may give an alternative definition of a differentiable manifold: it is a topological space M together with a subsheaf $\mathscr{C}^\infty(M) \subset \mathscr{C}(M)$ such that the pair $(M, \mathscr{C}^\infty(M))$ is locally isomorphic to an open subset of $\mathbf{R}^n$ with its sheaf of differentiable functions. Sheaves of functions can also be used to define many other kinds of geometric structures —for example, real analytic manifolds, complex analytic manifolds, and Nash manifolds may all be defined in this way. We will adopt an analogous approach in defining schemes: a scheme will be a topological space X with sheaf $\mathcal{O}_X$, locally isomorphic to an affine scheme as defined below.

Let R be a commutative ring. The affine scheme defined from R will be called Spec R, the *spectrum* of R. As indicated, it (like any scheme) consists of a set of points, a topology on it called the *Zariski topology*, and a sheaf $\mathcal{O}_{\mathrm{Spec}\,R}$ on this topological space, called the *sheaf of regular functions*, or *structure sheaf* of the scheme. Where there is a possibility of confusion we will use the notation $|\mathrm{Spec}\,R|$ to refer to the underlying set or topological space, without the sheaf, though if it is clear from context what we mean ("an open subset of $\mathrm{Spec}\,R$," for example), we may omit the vertical bars.

We will give the definition of the affine scheme $\mathrm{Spec}\,R$ in three stages, specifying first the underlying set, then the topological structure, and finally the sheaf.

i Schemes as Sets

We define a *point* of $\mathrm{Spec}\,R$ to be a prime—that is, a prime ideal—of R. To avoid confusion, we will sometimes write $[\mathfrak{p}]$ for the point of $\mathrm{Spec}\,R$ corresponding to the prime $\mathfrak{p}$ of R. We will adopt for this the usual convention that R itself is not a prime ideal. Of course, the zero ideal (0) is a prime if R is a domain.

In case R is the coordinate ring of an ordinary affine variety V over an algebraically closed field, $\mathrm{Spec}\,R$ will have points corresponding to the points of the affine variety—the maximal ideals of R—and also a point corresponding to each irreducible subvariety of V. The new points, corresponding to positive dimensional subvarieties, are at first rather unsettling but turn out to be quite convenient. They play the role of the "generic points" of classical algebraic geometry.

Each element $f \in R$ defines a sort of function on the space $\mathrm{Spec}\,R$: if $x = [\mathfrak{p}] \in \mathrm{Spec}\,R$, we write $\kappa(x)$ or $\kappa(\mathfrak{p})$ for the quotient field of the integral domain $R/\mathfrak{p}$, called the *residue field of X at x*, and we define $f(x) \in \kappa(x)$ to be the image of f via the canonical maps

$$R \to R/\mathfrak{p} \to \kappa(x)$$

We define a *regular function* on $\mathrm{Spec}\,R$ to be one of these functions, coming from an element of R.

EXERCISE I-1 ____

a. Consider the ring of polynomials $\mathbf{C}[x]$, and let $p(x)$ be a polynomial. Show that if $\alpha \in \mathbf{C}$ is a number, then $(x - \alpha)$ is a prime of $\mathbf{C}[x]$, and there is a natural identification of $\kappa((x - \alpha))$ with $\mathbf{C}$ such that the value of $p(x)$ at the point $(x - \alpha) \in \mathrm{Spec}\,\mathbf{C}[x]$ is the number $p(\alpha)$.

b. More generally, if R is the coordinate ring of an affine variety V over an algebraically closed field k, and $\mathfrak{p}$ is the maximal ideal corresponding to an ordinary point $x \in V$, then $\kappa(x) = k$ and $f(x)$ is the value of f at x in the usual sense.

Note that, in general, the "function" f has values in fields that vary from point to point. Moreover, f is not necessarily determined by the values of this function. For example, the ring $k[x]/(x^2)$ has only one prime ideal, which is (x); and thus the nonzero element $x \in R$ induces a function whose value is 0 at every point of Spec R, but that is not equal to zero.

EXERCISE I-2

Consider the ring of integers $\mathbf{Z}$. What is the value of the "function" 15 at the point $(7) \in \mathrm{Spec}(\mathbf{Z})$? At the point (5)?

ii Schemes as Topological Spaces

By using these functions, we make Spec R into a topological space; the topology is called the *Zariski topology*. The closed sets are defined as follows. For each subset $S \subset R$, let

$$V(S) = \{x \in \mathrm{Spec}\, R \mid f(x) = 0 \text{ for all } f \in S\} = \{[\mathfrak{p}] \in \mathrm{Spec}\, R \mid \mathfrak{p} \supset S\}$$

The impulse behind this definition is to make the functions $f \in R$ behave as much like continuous functions as possible. Of course, the fields $\kappa(x)$ have no topology, and since they vary with x, the usual notion of continuity makes no sense. But at least they all contain an element called zero, so one can speak of the locus of points in Spec R on which f is zero; and if f is to be like a continuous function, this locus should be closed. Since intersections of closed sets must be closed, we are led immediately to the definition above: $V(S)$ is just the intersection of the loci where the elements of S vanish.

For the family of sets $V(S)$ to be the closed sets of a topology, it is necessary that it be closed under arbitrary intersections; from the description above it is clear that for any family of sets S_a, we have $\bigcap_a V(S_a) = V(\bigcup_a S_a)$, as required. It is worth noting also that if I is the ideal generated by S, then $V(I) = V(S)$.

An open set in the Zariski topology is simply the complement of one of the $V(S)$. The open sets corresponding to sets S with just one element will play a special role, essentially because they are again spectra of rings; for this reason they get a special name and notation. If $f \in R$, then we define the *distinguished* (or *basic*) open subset of $X = \mathrm{Spec}\, R$ associated with f to be

$$X_f = |\mathrm{Spec}\, R| - V(f) = |\mathrm{Spec}(R_f)|$$

where R_f is the localization of R obtained by adjoining an inverse to f and the last equality comes from the fact that the prime ideals of R_f correspond to the prime ideals of R that do not contain f, by the correspondence that sends $\mathfrak{p} \subset R$ to $\mathfrak{p}R_f \subset R_f$. The distinguished open sets form a *base* for the Zariski topology in the sense that any open set is a union of distinguished ones:

$$U = \operatorname{Spec} R - V(S) = \operatorname{Spec} R - \bigcap_{f \in S} V(f) = \bigcup_{f \in S} (\operatorname{Spec} R)_f$$

They are also closed under finite intersections; since a prime ideal contains a product iff it contains one of the factors, we have

$$\bigcap_{i=1,\ldots,n} (\operatorname{Spec} R)_{f_i} = (\operatorname{Spec} R)_g$$

where g is the product $f_1 \cdots f_n$. In particular, any distinguished open set that is a subset of the distinguished open set $(\operatorname{Spec} R)_f$ has the form $(\operatorname{Spec} R)_{f \cdot g}$ for suitable g.

Spec R is almost never a Hausdorff space—the open sets are simply too large. In fact, the only points of Spec R that are closed are those corresponding to maximal ideals of R. In general, it is clear that the smallest closed set containing a given point $[\mathfrak{p}]$ must be $V(\mathfrak{p})$, so the closure of the point $[\mathfrak{p}]$ consists of all $[\mathfrak{q}]$ such that $\mathfrak{q} \supset \mathfrak{p}$. Thus in the case where R is the affine ring of an algebraic variety V over an algebraically closed field, the points of V correspond precisely to the closed points of Spec R, and the closed points contained in the closure of the point $[\mathfrak{p}]$ are exactly the points of V in the subvariety determined by $\mathfrak{p}$.

EXERCISE I-3

a. The points of $\operatorname{Spec}(\mathbf{C}[x])$ are the primes $(x - a)$, for every $a \in \mathbf{C}$, and the prime (0). Describe the topology. In particular, which points are closed? Are any of them open?

b. Let k be a field and let R be the local ring $k[x]_{(x)}$. Describe the topological space Spec R. (The answer is given later in this section.)

To complete the definition of Spec R, we have to describe the structure sheaf, or sheaf of regular functions on X. Before doing this, we will take a moment out to give some of the basic definitions of sheaf theory and to prove a lemma that will be essential later on.

iii An Interlude on Sheaf Theory

Let X be any topological space. A *presheaf* $\mathscr{F}$ on X assigns to each open set U in X a set, denoted $\mathscr{F}(U)$, and to every pair of nested open sets $U \subset V \subset X$ a *restriction map*

$$\mathrm{res}_{V,U} : \mathscr{F}(V) \to \mathscr{F}(U)$$

satisfying the basic properties that

$$\mathrm{res}_{U,U} = \text{identity}$$

and

$$\text{for all } U \subset V \subset W \subset X, \qquad \mathrm{res}_{V,U} \circ \mathrm{res}_{W,V} = \mathrm{res}_{W,U}$$

The elements of $\mathscr{F}(U)$ are called the *sections of $\mathscr{F}$ over U*; elements of $\mathscr{F}(X)$ are called *global sections*.

Another way to express this is to define a *presheaf* to be a contravariant functor from the category of open sets in X (with a morphism $U \to V$ for each containment $U \subseteq V$) to the category of sets. Changing the target category to abelian groups, say, we have the definition of a presheaf of abelian groups, and the same goes for rings, algebras, and so on.

One of the most important constructions of this type is that of a presheaf of modules $\mathscr{F}$ over a presheaf of rings $\mathcal{O}$ on a space X. Such a thing is a pair of:

for each open set U of X, a ring $\mathcal{O}(U)$ and an $\mathcal{O}(U)$-module $\mathscr{F}(U)$;

and

for each containment $U \supseteq V$, a ring homomorphism $\alpha : \mathcal{O}(U) \to \mathcal{O}(V)$ and a map of sets $\mathscr{F}(U) \to \mathscr{F}(V)$ that is a map of $\mathcal{O}(U)$-modules if we regard $\mathscr{F}(V)$ as an $\mathcal{O}(U)$-module by means of α.

A presheaf (of sets, abelian groups, rings, modules, and so on) is called a *sheaf* if it satisfies one further condition, called the *sheaf axiom*. This condition is that for each open covering $U = \bigcup_{a \in A} U_a$ of an open set $U \subset X$, and each collection of elements

$$f_a \in \mathscr{F}(U_a) \qquad \text{for each } a \in A$$

having the property that for all $a, b \in A$ the restrictions of f_a and f_b to $U_a \cap U_b$

are equal, there is a unique element $f \in \mathscr{F}(U)$ whose restriction to U_a is f_a for all a.

If $\mathscr{F}$ is a presheaf on X and U is an open subset of X, then we may define a presheaf $\mathscr{F}|_U$ on U, called the *restriction* of $\mathscr{F}$ to U, by setting $\mathscr{F}|_U(V) = \mathscr{F}(V)$ for any open subset V of U, the restriction maps being the same as those of $\mathscr{F}$ as well. It is easy to see that if $\mathscr{F}$ is actually a sheaf, then so is $\mathscr{F}|_U$.

In the sequel we shall work exclusively with presheaves and sheaves of things that are at least abelian groups, so we will usually omit the phrase "of abelian groups." Note that given two (pre-)sheaves of abelian groups, one can define their direct sum, tensor product, and so on, open set by open set; thus, for example, if $\mathscr{F}$ and $\mathscr{G}$ are sheaves of abelian groups, we define $\mathscr{F} \oplus \mathscr{G}$ by

$$\mathscr{F} \oplus \mathscr{G}(U) := \mathscr{F}(U) \oplus \mathscr{G}(U) \qquad \text{for any open set } U$$

The simplest sheaves on any topological space X are the sheaves of locally constant functions with values in a set K—that is, the sheaf $\mathscr{K}$, where $\mathscr{K}(U)$ is the set of locally constant functions from U to K; if K is a group, we may make $\mathscr{K}$ into a sheaf of groups by pointwise addition. Similarly, if K is a ring, and we define multiplication in $\mathscr{K}(U)$ to be pointwise multiplication, then $\mathscr{K}$ becomes a sheaf of rings. When K has a topology, we can define the *sheaf of continuous functions* with values in K (that is, the sheaf $\mathscr{C}$, where $\mathscr{C}(U)$ is the set of continuous functions from U to K, again with pointwise addition). If X is a differentiable manifold, there are also sheaves of differentiable functions, vector fields, differential forms, and so on. Generalizing all these examples, if $\pi : Y \to X$ is any map of topological spaces, then we may define the sheaf $\mathscr{S}$ of germs of sections of π; that is, for every open set U of X we define $\mathscr{S}(U)$ to be the set of continuous maps $\sigma : U \to \pi^{-1}U$ such that $\pi\sigma = 1$.

EXERCISE I-4

a. Let X be the two-element set $\{0, 1\}$, and make X into a topological space by taking each of the four subsets to be open. A sheaf on X is thus a collection of four sets with certain maps between them; describe these. (X is actually $\mathrm{Spec}(R)$ for some rings R; can you find one?)

b. Do the same in the case where the topology on X is defined by taking only the three subsets $\varnothing$, $\{0\}$, $\{0, 1\}$ to be open. Again, this space may be realized as $\mathrm{Spec}(R)$.

c. Let V be a vector bundle on a topological space X. Check that the sheaf of germs of sections of V is a sheaf of modules over the sheaf of germs of continuous functions on X. (Sheaves of modules in general may in this way be seen as generalized vector bundles.)

Another way to describe a sheaf is by its stalks. For any presheaf $\mathscr{F}$ and any point $x \in X$, we define the *stalk* $\mathscr{F}_x$ of $\mathscr{F}$ at x to be the *direct limit* of the groups $\mathscr{F}(U)$ over all open neighborhoods U of x in X—that is,

$$\mathscr{F}_x = \varinjlim_{U \text{ an open neighborhood of } x} \mathscr{F}(U)$$

$$:= \left\{ \begin{array}{l} \text{the disjoint union of } \mathscr{F}(U) \text{ over all open sets } U \text{ containing } x, \\ \text{modulo the equivalence relation } \sigma \sim \tau \text{ if } \sigma \in \mathscr{F}(U), \tau \in \mathscr{F}(V) \\ \text{and there is an open neighborhood } W \text{ of } x \text{ contained in } U \cap V \\ \text{such that the restrictions of } \sigma \text{ and } \tau \text{ to } W \text{ are equal—that is} \end{array} \right.$$

$$\mathrm{res}_{U,W}\, \sigma = \mathrm{res}_{V,W}\, \tau$$

For every $x \in U$ there is a map $\mathscr{F}(U) \to \mathscr{F}_x$ sending a section s to the equivalence class of (U,s); this class is denoted s_x. If $\mathscr{F}$ is a sheaf, a section $s \in \mathscr{F}(U)$ of $\mathscr{F}$ over U is determined by its images in the stalks $\mathscr{F}_x$ for all $x \in U$—equivalently, $s = 0$ iff $s_x = 0$ for all $x \in U$. This follows from the sheaf axiom: to say that $s_x = 0$ for all $x \in U$ is to say that for each x there is a neighborhood U_x of x in U such that $\mathrm{res}_{U,U_x}(s) = 0$, and then it follows that $s = 0$ in $\mathscr{F}(U)$.

This notion of stalks has a rather familiar geometric content: it is an abstraction of the notion of *germs*. For example, if X is an analytic manifold of dimension n, and $\mathcal{O}_X^{an}$ is the sheaf of analytic functions on X, then the stalk of $\mathcal{O}_X^{an}$ at x is the ring of germs of analytic functions at x—that is, the ring of convergent power series in n variables.

EXERCISE I-5

Find the stalks of the sheaves you produced for Exercise I-4.

EXERCISE I-6

Show that if we topologize the disjoint union $\overline{\mathscr{F}} = \bigcup \mathscr{F}_x$ by taking the images of the (U,s) to be a base for the open sets, then the natural map $\pi : \overline{\mathscr{F}} \to X$ is continuous, and for each U the set $\mathscr{F}(U)$ is just the set of continuous maps $s : U \to \overline{\mathscr{F}}$ such that $\pi \circ s$ is the identity on U. This contruction shows that any sheaf is the sheaf of germs of sections of a suitable map. In early works on sheaves they were defined this way. The topological space $\overline{\mathscr{F}}$ is called the "espace étalé" of the sheaf (because its open sets are "stretched out flat" over open sets of X).

A morphism $\varphi : \mathscr{F} \to \mathscr{G}$ of sheaves on a space X is defined simply to be a collection of maps $\varphi(U) : \mathscr{F}(U) \to \mathscr{G}(U)$ such that for every inclusion $U \subset V$ the diagram

$$\begin{array}{ccc} \mathscr{F}(V) & \xrightarrow{\ \varphi(V)\ } & \mathscr{G}(V) \\ {\scriptstyle \mathrm{res}_{V,U}}\big\downarrow & & \big\downarrow{\scriptstyle \mathrm{res}_{V,U}} \\ \mathscr{F}(U) & \xrightarrow{\ \varphi(U)\ } & \mathscr{G}(U) \end{array}$$

commutes. (In the categorical language, a morphism of sheaves is just a natural transformation of the corresponding functors from the category of open sets on X to the category of sets.) A morphism $\varphi : \mathscr{F} \to \mathscr{G}$ induces as well a map of stalks $\varphi_x : \mathscr{F}_x \to \mathscr{G}_x$ for each $x \in X$ and is determined by the data of the maps φ_x for all $x \in X$.

We say that a map $\varphi : \mathscr{F} \to \mathscr{G}$ of sheaves is injective, surjective, or bijective if each of the induced maps $\varphi_x : \mathscr{F}_x \to \mathscr{G}_x$ on stalks has the property. The following exercises show how this notion is related to the more naive notion, in terms of sections on arbitrary sets.

EXERCISE I-7

Show that if $\varphi : \mathscr{F} \to \mathscr{G}$ is a morphism of sheaves, then $\varphi(U)$ is injective (respectively, bijective) for all open sets $U \subset X$ if and only if φ_x is injective (respectively, bijective) for all points $x \in X$.

EXERCISE I-8

Show that Exercise I-7 is false if the condition "injective" is replaced by "surjective" by checking that in each of the following examples the maps induced by φ on stalks are surjective, but for some open set U the map $\varphi(U) : \mathscr{F}(U) \to \mathscr{G}(U)$ is not surjective.

1. Let X be the topological space $\mathbf{C} - \{0\}$, let $\mathscr{F} = \mathscr{G}$ be the sheaf of nowhere-zero, continuous, complex-valued functions, and let φ be the map sending a function f to f^2.

2. Let X be the Riemann sphere $\mathbf{C} \cup \{\infty\}$ and let $\mathscr{G}$ be the sheaf of analytic functions. Let $\mathscr{F}_1$ be the sheaf of analytic functions vanishing at 0; that is, $\mathscr{F}_1(U)$ is the set of analytic functions on U that vanish at 0 if $0 \in U$, and the set of all analytic functions on U if $0 \notin U$. Similarly, let $\mathscr{F}_2$ be the sheaf of analytic functions vanishing at ∞. Let $\mathscr{F} = \mathscr{F}_1 \oplus \mathscr{F}_2$, and let $\varphi : \mathscr{F} \to \mathscr{G}$ be the addition map.

These two examples are the beginning of the cohomology theory of sheaves; the reader will find more in this direction in the references on sheaves listed below.

If $\mathscr{F}$ is a presheaf on X, then we define the *sheafification* of $\mathscr{F}$ to be the unique sheaf $\mathscr{F}'$ and morphism of presheaves $\varphi : \mathscr{F} \to \mathscr{F}'$ such that for all $x \in X$ the map $\varphi_x : \mathscr{F}_x \to \mathscr{F}'_x$ is an isomorphism. More explicitly, the sheaf $\mathscr{F}'$ may be defined by saying that the sections of $\mathscr{F}'$ over an open set U are the functions s which take each point $x \in U$ to an element $s_x \in \mathscr{F}_x$ in such a way that s is locally induced by a section of $\mathscr{F}$; that is, such that there is an open cover of U by open sets U_i and elements $s_i \in \mathscr{F}(U_i)$ with $s_x = (s_i)_x$ for $x \in U_i$. The map $\mathscr{F} \to \mathscr{F}'$ is then gotten by sending $t \in \mathscr{F}(U)$ to the function

$x \mapsto t_x \in \mathscr{F}_x$ defined previously. The sheaf $\mathscr{F}'$ should be thought of as the sheaf "best approximating" the presheaf $\mathscr{F}$.

If $\varphi : \mathscr{F} \to \mathscr{G}$ is an injective map of sheaves, we will say that $\mathscr{G}$ is a *subsheaf* of $\mathscr{F}$. The notion of a quotient is more subtle. If $\mathscr{F}$ and $\mathscr{G}$ are presheaves of abelian groups, then their quotient as presheaves is the presheaf $\mathscr{H}$ defined by $\mathscr{H}(U) = \mathscr{G}(U)/\mathscr{F}(U)$. But if $\mathscr{F}$ and $\mathscr{G}$ are sheaves, then $\mathscr{H}$ will generally not be a sheaf, and we must define their *quotient* as sheaves to be the sheafification of $\mathscr{H}$, that is, $\mathscr{G}/\mathscr{F} := \mathscr{H}'$. The natural map from $\mathscr{H}$ to its sheafification $\mathscr{H}'$, together with the map of presheaves $\mathscr{G} \to \mathscr{H}$, defines the quotient map from $\mathscr{G}$ to $\mathscr{G}/\mathscr{F}$. This is the *cokernel* of φ.

The significance of the sheaf axiom is that sheaves are defined by local properties. We give two aspects of this principle explicitly.

In our applications to schemes, we will encounter a situation where we are given a *base* $\mathscr{B} = \{U_\alpha\}$ for the open sets of a topological space X, and we will want to specify a sheaf $\mathscr{F}$ just by saying what the groups $\mathscr{F}(U_\alpha)$ and homomorphisms $\mathrm{res}_{U_\beta, U_\alpha}$ are for open sets U_α of our base and inclusions $U_\alpha \subset U_\beta$ of basic sets. The main lemma is exactly the tool that says we can do this. To state it, we first make a definition: we say that a collection of groups $\mathscr{F}(U_\alpha)$ for open sets $U_\alpha \in \mathscr{B}$ and maps $\mathrm{res}_{U_\beta, U_\alpha} : \mathscr{F}(U_\beta) \to \mathscr{F}(U_\alpha)$ for $U_\alpha \subset U_\beta$ form a $\mathscr{B}$-*sheaf* if they satisfy the sheaf axioms with respect to inclusions of basic open sets in basic open sets and coverings of basic open sets by basic open sets.

PROPOSITION I-9

  i. Every $\mathscr{B}$-sheaf extends uniquely to a sheaf on X.

  ii. Given any two sheaves $\mathscr{F}$ and $\mathscr{G}$, and a collection of maps

$$\tilde{\varphi}(U_\alpha) : \mathscr{F}(U_\alpha) \to \mathscr{G}(U_\alpha) \qquad \text{for all } U_\alpha \in \mathscr{B}$$

commuting with restrictions, there is a unique morphism $\varphi : \mathscr{F} \to \mathscr{G}$ of sheaves such that $\varphi(U_\alpha) = \tilde{\varphi}(U_\alpha)$ for all $U_\alpha \in \mathscr{B}$.

Beginning of the Proof   For any open set $U \subset X$, set

$$\mathscr{F}(U) = \varprojlim_{U_\alpha \subset U, \, U_\alpha \in \mathscr{B}} \mathscr{F}(U_\alpha)$$

$$:= \left\{ \begin{array}{c} f = (f_\alpha \in U_\alpha) \in \prod_{U_\alpha \subset U, \, U_\alpha \in \mathscr{B}} \mathscr{F}(U_\alpha) \text{ such that} \\ \mathrm{res}_{U_\beta, U_\alpha}(f_\beta) = f_\alpha \text{ whenever } U_\alpha \subset U_\beta \subset U, \\ \text{with } U_\alpha \text{ and } U_\beta \in \mathscr{B} \end{array} \right\}$$

The restriction maps are defined immediately from the universal property of the inverse limit. $\qquad\qquad \square$

EXERCISE I-10

Complete the proof of the proposition by checking the sheaf axioms and showing that for $U_\alpha \in \mathscr{B}$, the new definition of $\mathscr{F}$ is the same as the old one.

The second application, which is really a special case of the first, says that to define a sheaf, it is enough to give it on each open set of an open cover, as long as the definitions are compatible.

COROLLARY I-11   Let $\{U_\alpha\}$ be an open covering of a topological space X. If $\mathscr{F}_\alpha$ is a sheaf on U_α for each α, and if

$$\varphi_{\alpha\beta} : \mathscr{F}_\alpha|_{U_\alpha \cap U_\beta} \to \mathscr{F}_\beta|_{U_\alpha \cap U_\beta}$$

are isomorphisms satisfying the compatibility conditions

$$\varphi_{\beta\gamma}\varphi_{\alpha\beta} = \varphi_{\alpha\gamma} \qquad \text{on } U_\alpha \cap U_\beta \cap U_\gamma$$

then there is a unique sheaf $\mathscr{F}$ on X whose restriction to each U_α is $\mathscr{F}_\alpha$.

Proof   The open sets contained in some U_α form a base $\mathscr{B}$ for the topology of X. For each such set V, we choose arbitrarily a set U_α that contains it, and define $\mathscr{F}(V) = \mathscr{F}_\alpha(V)$. If for some $W \subset V$ the value $\mathscr{F}(W)$ has been defined with reference to a different $\mathscr{F}_\beta$, then we use the isomorphism $\varphi_{\alpha\beta}$ to define the restriction maps. These compose correctly because of the compatibility conditions on the $\varphi_{\alpha\beta}$. Thus we have a $\mathscr{B}$-sheaf, and therefore a sheaf.   □

The pushforward operation on sheaves is so basic (and trivial) that we introduce it here: If $\alpha : X \to Y$ is a continuous map on topological spaces and $\mathscr{F}$ is a presheaf on X, we define the *pushforward* $\alpha_* \mathscr{F}$ of $\mathscr{F}$ by α to be the presheaf on Y given by

$$\alpha_* \mathscr{F}(V) := \mathscr{F}(\alpha^{-1}(V)) \qquad \text{for any open } V \subset Y$$

Of course, the pushforward of a sheaf of abelian groups (rings, modules over a sheaf of rings, and so on) is again of the same type.

EXERCISE I-12

Show that the pushforward of a sheaf is again a sheaf.

References for the theory of sheaves.   Serre's landmark paper "Faisceaux Algébrique Cohérent" (1955), which established sheaves as an important tool in algebraic geometry, is still a wonderful source of information. The books of Godement (1964) and Swan (1964) are more systematic introductions. Chapter II of the book of Hartshorne (1977) contains an excellent account adapted to the technical requirements of scheme theory; it is a simplified version of that found in EGA (1960–1967). Some good references for the analytic case are Forster (1981) (especially for an introduction to cohomology) and Gunning (1990, Vol. III).

iv Schemes as Schemes (Structure Sheaves)

We return at last to the definition of the scheme $X = \operatorname{Spec} R$. We will complete the construction by specifying the structure sheaf $\mathcal{O}_X = \mathcal{O}_{\operatorname{Spec} R}$. As indicated above, we want the relationship between $\operatorname{Spec} R$ and R to generalize that between an affine variety and its coordinate ring; in particular, we want the ring of global sections of the structure sheaf $\mathcal{O}_X$ to be R.

We thus wish to extend the ring R of functions on X to a whole sheaf of rings. This means that for each open set U of X, we wish to give a ring $\mathcal{O}_X(U)$; and for every pair of open sets $U \subset V$, we wish to give a restriction homomorphism

$$\operatorname{res}_{V,U} : \mathcal{O}_X(V) \to \mathcal{O}_X(U)$$

satisfying the various axioms above. It is quite easy to say what the rings $\mathcal{O}_X(U)$ and the maps $\operatorname{res}_{V,U}$ should be for distinguished open sets U and V: we set

$$\mathcal{O}_X(X_f) = R_f$$

and define $\operatorname{res}_{X_f, X_{fg}}$ to be the natural localization map $R_f \to R_{fg}$. Now, by Proposition I-9 on page 13 this will suffice to define the structure sheaf $\mathcal{O}_X$, as long as we verify that it satisfies the sheaf axiom with respect to coverings of distinguished opens by distinguished opens. Before doing this, in Proposition I-15 below, we exhibit a simple but fundamental lemma that describes the coverings of affine schemes by distinguished open sets.

LEMMA I-13   Let $X = \operatorname{Spec} R$, and let $\{f_a\}$ be a collection of elements of R. The open sets X_{f_a} cover X iff the elements f_a generate the unit ideal. In particular, X is quasicompact as a topological space.

Recall that *quasicompact* means that every open cover has a finite subcover; the *quasi* is there because the space is not necessarily Hausdorff. In fact, schemes are almost never Hausdorff! Unfortunately, this fact vitiates most of the usual advantages of compactness. For example, in contrast to the situation for compact manifolds, say, the continuous image of one affine scheme in another need not be closed. For this reason, we will discuss in Section III-A a better "compactness" notion, called *properness*, which will play just as important a role as compactness does in the usual geometric theories.

Proof   The X_{f_a} cover X iff no prime of R contains all of the f_a iff the f_a generate the unit ideal; this proves the first statement. To prove the second statement, note first that every open cover $X = \bigcup X_b$ has a refinement of the form $X = \bigcup X_{f_a}$, where $f_a \in R$ and each X_{f_a} is contained in some X_b. The X_{f_a} cover X iff the f_a generate the unit ideal. This in turn will be the case iff the element 1 can be written as a linear combination—necessarily finite—of the f_a. Taking just the f_a involved in this expansion of 1, we see that the cover $X = \bigcup X_{f_a}$, and with it the original cover $X = \bigcup X_b$, has a finite subcover.

$\square$

EXERCISE I-14 ____

If R is Noetherian, then *every* subset of Spec R is quasicompact.

PROPOSITION I-15   Let $X = \text{Spec } R$. If X_f is covered by open sets X_{f_a}, then

  a. if $g, h \in R_f$ become equal in each R_{f_a}, then they are equal;
  b. if for each a we have $g_a \in R_{f_a}$ such that for each pair a and b the images of g_a and g_b in $R_{f_a f_b}$ are equal, then there is an element $g \in R_f$ whose image in R_{f_a} is g_a for all a.

Equivalently, if $\mathscr{B}$ is the collection of distinguished open sets Spec R_f of Spec R, and if we set $\mathcal{O}_X(\text{Spec } R_f) := R_f$, then $\mathcal{O}_X$ is a $\mathscr{B}$-sheaf. By Proposition I-9, $\mathcal{O}_X$ extends uniquely to a sheaf on X.

DEFINITION I-16   The sheaf $\mathcal{O}_X$ defined in the proposition is called the *structure sheaf* X, or the *sheaf of regular functions on X*.

PROOF OF PROPOSITION I-15   To simplify the notation, we may assume that $R_f = R$ and $X_f = X$.

  a. If g and h become equal in each X_{f_a} then $g - h$ is annihilated by a power of each f_a. Since by Lemma I-13 we may assume that the cover is finite, this implies that $g - h$ is annihilated by a power of the ideal generated by all the f_a^N for some N. But this ideal contains a power of the ideal generated by all the f_a, which is the unit ideal. Thus $g = h$ in R.

  b. We will use an argument analogous to the classical partition of unity to piece together the elements g_a into a single element $g \in R$.

Since g_a and g_b become equal on $X_{f_a f_b}$ we must have

$$(f_a f_b)^N g_a = (f_a f_b)^N g_b$$

for all large N, and since by Lemma I-13 we may assume that the covering $\{X_{f_a}\}$ is finite, one N will do for all a, b. Again by Lemma I-13 the elements f_a, and with them the elements f_a^N, generate the unit ideal, so we may write

$$1 = \sum_a e_a f_a^N \qquad \text{for some } e_a \in R;$$

this is our "partition of unity." We claim that

$$g = \sum_a e_a f_a^N g_a$$

is the element we seek. Indeed, for each b,

$$f_b^N g = \sum_a f_b^N e_a f_a^N g_a = \sum_a f_b^N e_a f_a^N g_b$$

$$= f_b^N \left(\sum_a e_a f_a^N \right) g_b = f_b^N g_b$$

so g becomes equal to g_b on X_{f_b} as required. $\qquad\qquad\square$

The proposition is still valid (and has essentially the same proof) if we replace R_f and R_{f_α} by M_f and M_{f_α} for any R-module M.

A trivial but occasionally confusing point deserves a remark. The empty set $\varnothing$ is of course an open subset of $\operatorname{Spec} R$; the general mechanism for defining sheaves mentioned above leads to

$$\mathcal{O}_X(\varnothing) = 0$$

the "zero ring" consisting of only one element 0 (which is also the unit element!). Note that the zero ring has no prime ideals at all—it is the only ring with unit having this property, if one accepts the axiom of choice—so that indeed $\operatorname{Spec} 0 = \varnothing$, and everything is consistent.

EXERCISE I-17

Describe the points and the sheaf of functions of each of the following schemes.

i. $X_1 = \operatorname{Spec} \mathbf{C}[x]/(x^2)$

ii. $X_2 = \operatorname{Spec} \mathbf{C}[x]/(x^2 - x)$

iii. $X_3 = \operatorname{Spec} \mathbf{C}[x]/(x^3 - x^2)$

iv. $X_4 = \operatorname{Spec} \mathbf{R}[x]/(x^2 + 1)$

In contrast with the situation in many geometric theories (though similar to the situation in the category of complex manifolds), there may be really rather "few" regular functions on a scheme. For example, when we define arbitrary schemes, we shall see that the schemes that are the analogues of compact manifolds may have no nonconstant regular functions on them at all. For this reason, partially defined functions on a scheme X (that is, elements of $\mathcal{O}_X(U)$ for some open dense subset U) play an unusually large role. They are called *rational* functions on X because in the case $X = \operatorname{Spec} R$ with R a domain, and $U = X_f$, the elements of $\mathcal{O}_X(X_f) = R_f$ are ratios of elements in R. In the cases of most interest, we shall see that every nonempty open set is dense in X, so the behavior of rational functions reflects the properties of X as a whole.

EXERCISE I-18* ____

Compute the *ring of rational functions*

$$\lim_{\substack{\longrightarrow \\ U \text{ an open set of } X}} \mathcal{O}_X(U)$$

$$:= \left\{ \begin{array}{c} \text{the disjoint union of } \mathcal{O}_X(U) \text{ for all open dense sets} \\ U \text{ modulo the equivalence relation} \\ \sigma \sim \tau \quad \text{if} \quad \sigma \in \mathcal{O}_X(U), \quad \tau \in \mathcal{O}_X(V) \\ \text{and the restrictions of } \sigma \text{ and } \tau \text{ are equal} \\ \text{on some open dense } W \subset U \cap V \end{array} \right.$$

first in the case where R is a domain and then for an arbitrary Noetherian ring.

Example I-19   Another very simple example will perhaps help to fix these ideas. Let k be a field, and let $R = k[x]_{(x)}$, the localization of the polynomial ring in one variable x at the maximal ideal (x). (*Note:* This is different from the ring $k[x]_x = k[x, x^{-1}]$!) The scheme $X = \operatorname{Spec} R$ has only two points, corresponding to the two prime ideals 0 and (x) of R. As a topological space, it has precisely three open sets,

$$\varnothing \subset U := \{[0]\} \subset \{[0], [(x)]\} = X$$

U and ϕ are both distinguished open sets, since $\{[0]\} = X_x$. The sheaf $\mathcal{O}_X$ is thus easy to describe: it has values

$$\mathcal{O}_X(X) = R = k[x]_{(x)}$$

$$\mathcal{O}_X(U) = R_x = k(x), \qquad \text{the field of rational functions}$$

and the restriction map from the first to the second is the natural inclusion.

EXERCISE I-20

Give a similarly complete description for the structure sheaf of the scheme $\operatorname{Spec} k[x]$. (The answer is given in Chapter II.)

B Schemes in General

After this lengthy description of affine schemes, the general definition of a scheme may be an anticlimax. A *scheme* X is defined to be a topological space, called the *support* of X and denoted $|X|$ or $\operatorname{supp} X$, together with a sheaf $\mathcal{O}_X$ of rings, such that the pair $(|X|, \mathcal{O}_X)$ is locally isomorphic to an affine scheme —that is, for every point $p \in X$ there exists a neighborhood U of p in X and a homeomorphism ψ of U to an affine scheme $Y = \operatorname{Spec} R$ such that $\psi_*(\mathcal{O}_X|_U) \cong \mathcal{O}_Y$, where $\psi_*(\mathcal{O}_X|_U)$ is the sheaf given by

$$\psi_*(\mathcal{O}_X|_U)(W) = \mathcal{O}_X(\psi^{-1}W) \qquad \text{for all open sets } W \subset Y$$

As we indicated before, where there is no danger of confusion, we will use the same letter X to denote the scheme and the underlying space $|X|$, as in the construction "let $p \in X$ be a point."

EXERCISE I-21

The smallest nonaffine scheme: let X be the topological space with three points $p, q_1,$ and q_2. Topologize X by making $X_1 := \{p, q_1\}$ and $X_2 := \{p, q_2\}$ open sets (so that, in addition, $\varnothing$, $\{p\}$, and X itself are open). Define a presheaf $\mathcal{O}$ of rings on X by setting

$$\mathcal{O}(X) = \mathcal{O}(X_1) = \mathcal{O}(X_2) = k[x]_{(x)}, \qquad \mathcal{O}(\{p\}) = k(x)$$

with restriction maps $\mathcal{O}(X) \to \mathcal{O}(X_i)$ the identity, and $\mathcal{O}(X_i) \to \mathcal{O}(\{p\})$ the obvious inclusion. Check that this presheaf is a sheaf and that $(X, \mathcal{O})$ is a scheme. Show that it is not an affine scheme. (Geometrically, the scheme

$(X, \mathcal{O})$ is the "germ of the doubled point" in the scheme called X_1 in Exercise I-41 on page 31.)

i Closed and Open Subschemes

Consider first an affine scheme $X = \operatorname{Spec} R$. For any ideal I in the ring R, we may make the closed subset $V(I) \subset X$ into an affine scheme by identifying it with $Y := \operatorname{Spec} R/I$. This makes sense because the primes of R/I are exactly the primes of R that contain I, taken modulo I, and thus the topological space $|\operatorname{Spec} R/I|$ is canonically homeomorphic to the closed set $V(I) \subset X$. We define a *closed subscheme* Y of X to be a scheme of this form, together with the identification of Y with the spectrum of a quotient ring of R (so that the closed subschemes of X by definition correspond one to one with the ideals in the ring R).

We can define in these terms all the usual operations on and relations between subschemes of a given scheme $X = \operatorname{Spec} R$. Thus, we say that the closed subscheme $Y = \operatorname{Spec} R/I$ of X *contains* the closed subscheme $Z = \operatorname{Spec} R/J$ if Z is in turn a closed subscheme of Y—that is, if $J \supset I$. We note that this implies that $V(J) \subset V(I)$ but not conversely. The *union* and *intersection* of the closed subschemes $\operatorname{Spec} R/I$ and $\operatorname{Spec} R/J$ are by definition $\operatorname{Spec} R/(I \cap J)$ and $\operatorname{Spec} R/(I + J)$. It is important to note that the notions of containment, intersection, and union do not satisfy all the usual axioms: for example, we will see in Section II-C-iii-a (page 62) an example of closed subschemes X, Y, Z of a scheme such that $X \cup Y = X \cup Z$ and $X \cap Y = X \cap Z$ but $Y \neq Z$.

We would like now to generalize the notion of closed subscheme to an arbitrary scheme X. To do this, the first step must be to replace the ideal $I \subset R$ associated to a subscheme Y of an affine scheme $X = \operatorname{Spec} R$ by a sheaf, which we do as follows. The structure sheaf $\mathcal{O}_Y$ is given on a distinguished open set $U = \operatorname{Spec}(R/I)_f$ by $\mathcal{O}_Y(U) = (R/I)_f = R_f/I_f$. We define $\mathscr{I} = \mathscr{I}_{Y/X}$, the *ideal sheaf of Y in X*, to be the sheaf of ideals of $\mathcal{O}_X$ given on a distinguished open set $V = X_g$ of X by $\mathscr{I}(X_g) = I \cdot R_g$. With this notation we may write $\mathcal{O}_Y$ or, more precisely, the pushforward $j_* \mathcal{O}_Y$, where j is the inclusion map $|Y| \subset |X|$, as $\mathcal{O}_X/\mathscr{I}$. The sheaf of ideals $\mathscr{I}$ may be recovered as the kernel of the restriction map $\mathcal{O}_X \to j_* \mathcal{O}_Y$.

There is a subtle point here that requires mention: not all sheaves of ideals in $\mathcal{O}_X$ arise from ideals of R. For example, in the case of $R = k[x]_{(x)}$ considered in Example I-19, we may define a sheaf of ideals $\mathscr{J}$ by

$$\mathscr{J}(U) = \mathcal{O}_X(U)$$

$$\mathscr{J}(X) = 0$$

For a sheaf of ideals $\mathscr{I}$ coming from an ideal of R we would have

$$\mathscr{I}(U) = \mathscr{I}(X)_x = \mathscr{I}(X) \cdot k(x)$$

so we see that $\mathscr{J}$ does not come from any ideal of R. In the definition of closed subscheme above, we are only interested in sheaves of ideals that do come from ideals of R. The theory obviously needs a name for these: they are called *quasicoherent* sheaves of ideals. (This seems a poor name for such a basic and simple object, but it is firmly rooted in the literature. It comes from the fact that a sheaf corresponding to a finitely generated module has a property that had previously been called coherence; it was thus natural to say that the sheaf coming from a finitely generated module is *coherent*, and that coming from an arbitrary module is quasicoherent.)

EXERCISE I-22

The schemes X_1, X_2, and X_3 of Exercise I-17 may all be viewed as subschemes of Spec $\mathbf{C}[x]$. Show that we have the inclusions

$$X_1 \subset X_3 \qquad \text{and} \qquad X_2 \subset X_3$$

but that no other inclusions $X_i \subset X_j$ hold, even though the underlying sets of X_2 and X_3 coincide and the underlying set of X_1 is contained in the underlying set of X_2.

We can generalize the notion of a closed subscheme of an affine scheme to the general case as follows. First, we define the notion of a *quasicoherent sheaf of ideals* $\mathscr{I} \subset \mathcal{O}_X$ on an arbitrary scheme X: this is just a sheaf of ideals $\mathscr{I}$ such that for every open affine subset U of X the restriction $\mathscr{I}|_U$ is a quasicoherent sheaf of ideals on U. If X is an arbitrary scheme, we may then make the following definition.

DEFINITION I-23   A *closed subscheme* of a scheme X is a closed topological subspace Y together with a sheaf of rings $\mathcal{O}_Y$ that is a quotient sheaf of the structure sheaf $\mathcal{O}_X$ by a quasicoherent sheaf of ideals $\mathscr{I}$, such that the intersection of Y with any affine open subset $U \subset X$ is the closed subscheme associated to the ideal $\mathscr{I}(U)$.

In other words, a closed subscheme is something that looks locally like a closed subscheme of an affine scheme. Note that the closed subschemes of X are in one-to-one correspondence with the quasicoherent sheaves of ideals $\mathscr{I} \subset \mathcal{O}_X$.

The notion of quasicoherence arises in a more general context as well. We similarly define a *quasicoherent sheaf* $\mathscr{F}$ on X to be a sheaf of $\mathcal{O}_X$ modules (that is, $\mathscr{F}(U)$ is an $\mathcal{O}_X(U)$-module for each U) such that for any affine set U and distinguished open subset $U_f \subset U$, the $\mathcal{O}_X(U_f) = \mathcal{O}_X(U)_f$-module $\mathscr{F}(U_f)$ is obtained from $\mathscr{F}(U)$ by inverting f—more precisely, the restriction map $\mathscr{F}(U) \to \mathscr{F}(U_f)$ becomes an isomorphism after inverting f. $\mathscr{F}$ is called *coherent* if all the modules $\mathscr{F}(U)$ are finitely generated. One might say informally that these are the sheaves of modules whose restrictions to open affine sets "are" modules (respectively finitely generated modules) on the corresponding rings. The notion of quasicoherent sheaf is the right analogue in the context of schemes of the notion of module over a ring; for most purposes, one should think of them simply as modules.

EXERCISE I-24*

To check that a sheaf of ideals (or any sheaf of modules) is quasicoherent (or for that matter coherent), it is enough to check the defining property on each open affine set U of a fixed affine cover of X.

One of the most important subschemes of an affine scheme X is X_{red}, the *reduced scheme associated to* X. This may be defined by setting $X_{\mathrm{red}} = \operatorname{Spec} R_{\mathrm{red}}$, where R_{red} is R modulo its *nilradical*—that is, modulo the ideal of nilpotent elements of R. Since any prime $\mathfrak{p} \subset R$ contains all nilpotent elements (in fact, the set of nilpotent elements is the intersection of the minimal primes), $|X|$ and $|X_{\mathrm{red}}|$ are identical as topological spaces.

EXERCISE I-25

X_{red} may also be defined as the topological space $|X|$ with structure sheaf $\mathcal{O}_{X_{\mathrm{red}}}$ associating to every open subset $U \subset X$ the ring $\mathcal{O}_X(U)$ modulo its nilradical.

To globalize this notion, we may define for any scheme X a sheaf of ideals $\mathcal{N} \subset \mathcal{O}_X$, called the *nilradical*; this is the sheaf whose value on any open set U is the nilradical of $\mathcal{O}_X(U)$. The associated closed subscheme of X is called the reduced scheme associated to X and denoted X_{red}.

By way of vocabulary, we say that a scheme X is *reduced* if $X = X_{\mathrm{red}}$, and it is *irreducible* if the underlying topological space of X is not the union of two properly smaller closed sets. Here are some easy but important remarks about these notions.

EXERCISE I-26

Show that an affine scheme $X = \operatorname{Spec} R$ is reduced and irreducible iff R is a domain. Then show that it is irreducible iff R has a unique minimal prime.

EXERCISE I-27 ____

Show that a scheme X is reduced iff every open affine subscheme of X is reduced iff every local ring $\mathcal{O}_{X,p}$ is reduced for closed points $p \in X$.

EXERCISE I-28 ____

Show that a topological space is irreducible iff every open subset is dense.

EXERCISE I-29 ____

Show that the disjoint union of two affine schemes $\operatorname{Spec} R$ and $\operatorname{Spec} S$ may be identified with the scheme $\operatorname{Spec} R \times S$.

EXERCISE I-30 ____

Now show that an arbitrary scheme X is irreducible iff every open affine subset is irreducible. If it is connected (in the sense that the topological space $|X|$ is connected), then it is irreducible iff every local ring has a unique minimal prime.

Now let U be an open subset of a scheme X. The pair $(U, \mathcal{O}_X|_U)$ is again a scheme, but this requires an argument—for example, it need not be an affine scheme even if X is affine.

EXERCISE I-31* ____

Show that the complement U of the point $[(x, y)]$ in $\operatorname{Spec} k[x, y]$ is not affine.

To see that $(U, \mathcal{O}_X|_U)$ is really a scheme, note that at least a *distinguished* open set of an affine scheme is again an affine scheme: if $X = \operatorname{Spec} R$, and $U = X_f$, then $(U, \mathcal{O}_X|_U) = \operatorname{Spec} R_f$. Since the distinguished open sets of X that are contained in U cover U, this shows that that $(U, \mathcal{O}_X|_U)$ is covered by affine schemes, as required. An open subset of a scheme is correspondingly sometimes referred to as an *open subscheme* of X, with this structure understood. Note that in contrast to the case of a closed subscheme, an open subscheme is determined by its underlying subset $|U| \subset |X|$.

The Noetherian property is fundamental in the theory of rings, and its extension is equally fundamental in the theory of schemes: we say that a scheme X is *Noetherian* if it admits a finite cover by open affine subschemes, each the spectrum of a Noetherian ring. As usual, one can check that this is independent of the cover chosen.

There is a good notion of the germ of a scheme X at a point $x \in X$, which is the intersection, in a natural sense, of all the open subschemes containing the point. This is embodied in the *local ring* of X at x, defined earlier as

$$\mathcal{O}_{X,x} := \varinjlim_{x \in U} \mathcal{O}_X(U)$$

The maximal ideal $\mathfrak{m}_{X,x}$ of this local ring is given by the set of all those sections vanishing at x. This is a simple object: to compute it (and to show in particular that it *is* a local ring, with the given maximal ideal), we may begin by replacing X by an affine open neighborhood of x, and thus assume that $X = \operatorname{Spec} R$, $x = [\mathfrak{p}]$. We may next restrict the open subsets U in the $\varinjlim$ to the distinguished open sets $\operatorname{Spec} R_f$ such that $f(x) \neq 0$—that is, $f \notin \mathfrak{p}$. Thus

$$\mathcal{O}_{X,x} = \varinjlim_{f \notin \mathfrak{p}} R_f = R_{\mathfrak{p}}$$

$$\mathfrak{m}_{X,x} = \varinjlim_{f \notin \mathfrak{p}} \mathfrak{p} R_f = \mathfrak{p} R_{\mathfrak{p}}$$

the localization of R at $\mathfrak{p}$. We can think of the germ of X at x as being $\operatorname{Spec} \mathcal{O}_{X,x}$; we will study some schemes of this type in the examples of the next chapter.

This notion of the local ring of a scheme at a point is crucial to the whole theory of schemes. We give a few illustrations, showing how to define various geometric notions in terms of the local ring. Let X be a scheme.

1. The *dimension of X at a point* $x \in X$, written $\dim(X, x)$, is the (Krull) dimension of the local ring $\mathcal{O}_{X,x}$—that is, the supremum of lengths of chains of prime ideals in $\mathcal{O}_{X,x}$. The *dimension* of X, or $\dim X$, itself is the supremum of these local dimensions.

EXERCISE I-32

The underlying space of a zero-dimensional scheme is a discrete set of points. If the scheme is Noetherian, the set is finite.

2. The *Zariski cotangent space* to X at x is $\mathfrak{m}_{X,x}/(\mathfrak{m}_{X,x})^2$, regarded as a vector space over the residue field $\kappa(x) = \mathcal{O}_{X,x}/\mathfrak{m}_{X,x}$. The dual of this vector space is called the *Zariski tangent space* at x.

To understand this definition, consider first a complex algebraic variety X that is nonsingular. In this setting the notion of the tangent space to X at a point p is unambiguous: it may be taken as the vector space of derivations from the ring of germs of analytic functions at the point into $\mathbf{C}$. If $\mathfrak{m}_{X,p}$ is the ideal of regular functions vanishing at p, then such a derivation induces a $\mathbf{C}$-linear map $\mathfrak{m}_{X,p}/\mathfrak{m}_{X,p}^2 \to \mathbf{C}$, and the tangent space may be identified in this way with $\mathrm{Hom}_{\mathbf{C}}(\mathfrak{m}_{X,p}/\mathfrak{m}_{X,p}^2, \mathbf{C}) = (\mathfrak{m}_{X,p}/\mathfrak{m}_{X,p}^2)^*$. It was Zariski's insight that this latter vector space is the correct analogue of the tangent space for any point, smooth or singular, on any variety; Grothendieck subsequently carried the idea over to the context of schemes, as above. We shall return to this construction, from a new point of view, in Chapter IV.

EXERCISE I-33

If k is a field, the Zariski tangent space to $\mathrm{Spec}\, k[x_1, \ldots, x_n]$ at $x = [(x_1, \ldots, x_n)]$ is n-dimensional.

3. X is said to be *nonsingular* (or *regular*) at $x \in X$ if the Zariski tangent space to X at x has dimension equal to $\dim(X, x)$; else the dimension of the Zariski tangent space must be larger, and we say that X is *singular* at x. Thus in the case of primary interest, when X is Noetherian, X is nonsingular at x iff the local ring $\mathcal{O}_{X,x}$ is a regular local ring. This fundamental notion represents, historically, one of the important steps toward the algebraization of geometry. It was taken by Zariski in his classic paper (1947) (remarkably, this was some years *after* Krull had introduced the notion of a regular local ring to generalize the properties of polynomial rings, one of the rare cases in which the algebraists beat the geometers to a fundamental geometric notion).

EXERCISE I-34

A zero-dimensional Noetherian scheme is nonsingular iff it is the union of reduced points.

ii Morphisms

We will next define morphisms of schemes. In the classical theory a regular map of affine varieties gives rise, by composition, to a map of coordinate rings going in the opposite direction. This correspondence makes the two kinds of objects (regular maps of affine varieties and algebra homomorphisms of their coordinate algebras) equivalent. The definition given below generalizes this: we will see that maps between affine schemes are simply given by maps of the corresponding rings (in the opposite direction).

Given the simple description of morphisms of affine schemes in terms of maps of rings, it is tempting just to define a morphism of schemes to be something that is "locally a morphism of affine schemes." One can make sense of this, and it gives the correct answer, but it leads to awkward problems of checking that the definition is independent of the choice of an affine cover. For this reason, we give a definition below that works without the choice of an affine cover. Although it may at first appear complicated, it is quite convenient in practice. It also has the advantage of working uniformly for all "local ringed spaces"—structures defined by a topological space with a sheaf of rings whose stalks are local rings.

To understand the motivation behind this definition, consider once more the case of differentiable manifolds. A continuous map $\psi : M \to N$ between differentiable manifolds is differentiable if and only if for every differentiable function f on an open subset $U \subset N$, the pullback $\psi^*(f) := f \circ \psi$ is a differentiable function on $\psi^{-1}(U) \subset M$. We can express this readily enough in the language of sheaves: a differentiable map $\psi : M \to N$ may be defined as a continuous map $\psi : M \to N$ such that the natural map ψ^* on continuous functions induces a map, also called ψ^*, from the subsheaf $\mathscr{C}^\infty(N) \subset \mathscr{C}(N)$ to the subsheaf $\psi_* \mathscr{C}^\infty(M) \subset \psi_* \mathscr{C}(M)$. That is, we require that there be a commutative diagram

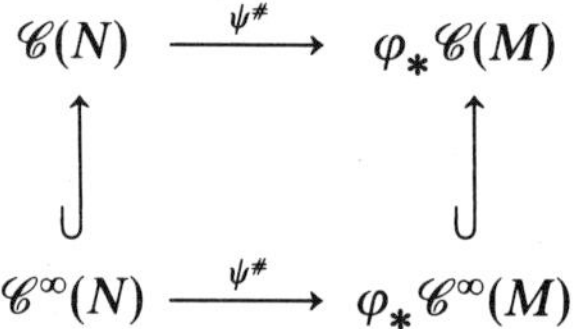

$$
\begin{array}{ccc}
\mathscr{C}(N) & \xrightarrow{\ \psi^\#\ } & \varphi_* \mathscr{C}(M) \\
\big\uparrow & & \big\uparrow \\
\mathscr{C}^\infty(N) & \xrightarrow{\ \psi^\#\ } & \varphi_* \mathscr{C}^\infty(M)
\end{array}
$$

We would like to adapt this idea to the case of schemes. The difference is that the structure sheaf $\mathcal{O}_X$ of a scheme X is *not* a subsheaf of a predefined sheaf of functions on X. Thus in order to give a map of schemes, we have to specify *both* a continuous map $\psi : X \to Y$ of underlying topological spaces *and* a pullback map

$$
\psi^\# : \mathcal{O}_Y \to \psi_* \mathcal{O}_X
$$

Of course, the map $\psi^\#$ has to satisfy some compatibility conditions with the map ψ. The problem in specifying these is that a section of the structure sheaf $\mathcal{O}_Y$ does not take values in a fixed field but in a field $\kappa(q)$ that varies with the point $q \in Y$; in particular, it doesn't make sense to require that the value of $f \in \mathcal{O}_Y(U)$ at $q \in U \subset Y$ agree with the value of $\psi^\#(f) \in \psi_* \mathcal{O}_X(U) = \mathcal{O}_X(\psi^{-1} U)$ at a point $p \in \psi^{-1} U \subset X$ mapping to q, which is in effect how $\psi^\#$ was defined in the case of differentiable functions. About all that does make sense is to require that f vanish at q if and only if $\psi^\# f$ vanishes at p—and this is exactly what we do require. We thus make the following definition.

DEFINITION I-35   A *morphism*, or *map*, between schemes X and Y is a pair $(\psi, \psi^{\#})$, where $\psi : X \to Y$ is a continuous map on the underlying topological spaces and

$$\psi^{\#} : \mathcal{O}_Y \to \psi_* \mathcal{O}_X$$

a map of sheaves on Y satisfying the condition that for any point $p \in X$ and any neighborhood U of $q = \psi(p)$ in Y, a section $f \in \mathcal{O}_Y(U)$ vanishes at q if and only if the section $\psi^{\#}(f)$ of $\psi_* \mathcal{O}_X(U) = \mathcal{O}_X(\psi^{-1}U)$ vanishes at p.

This last condition has a nice reformulation in terms of the local rings $\mathcal{O}_{X,p}$ and $\mathcal{O}_{Y,q}$. Any map of sheaves $\psi^{\#} : \mathcal{O}_Y \to \psi_* \mathcal{O}_X$ induces on passing to the limit a map

$$\mathcal{O}_{Y,q} = \varinjlim_{q \in U} \mathcal{O}_Y(U) \to \varinjlim_{q \in U} \mathcal{O}_X(\psi^{-1}U)$$

and this last ring naturally maps to the limit

$$\varinjlim_{p \in V} \mathcal{O}_X(V)$$

over all open subsets V containing p, which is $\mathcal{O}_{X,p}$. Thus $\psi^{\#}$ induces a map of the local rings $\mathcal{O}_{Y,q} \to \mathcal{O}_{X,p}$. The compatibility condition that a section $f \in \mathcal{O}_Y(U)$ vanishes at q if and only if the section $\psi^{\#}(f)$ of $\psi_* \mathcal{O}_X(U) = \mathcal{O}_X(\psi^{-1}U)$ vanishes at p simply says that this map sends the maximal ideal $\mathfrak{m}_{Y,q}$ into $\mathfrak{m}_{X,p}$; or as it is sometimes phrased, that $\psi^{\#}$ induces a *local* homomorphism of local rings.

As we mentioned above, a morphism of affine schemes

$$\psi : X = \operatorname{Spec} S \to \operatorname{Spec} R = Y$$

is the same as a homomorphism of rings $\varphi : R \to S$. Here is the precise result, along with an important improvement that describes maps from an arbitrary scheme to an affine scheme.

THEOREM I-36   For any scheme X and any ring R, the morphisms

$$(\psi, \psi^{\#}) : X \to \operatorname{Spec}(R)$$

are in one-to-one correspondence with the homomorphisms of rings

$$\varphi : R \to \mathcal{O}_X(X)$$

by the association

$$\varphi = \psi^{\#}(\mathrm{Spec}(R)) : R = \mathcal{O}_{\mathrm{Spec}(R)}(\mathrm{Spec}(R)) \to \psi_{*}(\mathcal{O}_X)(\mathrm{Spec}(R)) = \mathcal{O}_X(X)$$

Proof We describe the inverse association. Set $Y = \mathrm{Spec}(R)$, and let $\varphi : R \to \mathcal{O}_X(X)$ be a map of commutative rings. If $p \in |X|$ is a point, then the preimage of the maximal ideal under the composite $R \to \mathcal{O}_X(X) \to \mathcal{O}_{X,p}$ is a prime ideal, so that φ induces a map of sets

$$\psi : |X| \to |Y|$$

which is immediately seen to be continuous in the Zariski topology. Next, for each basic open set $U = \mathrm{Spec}(R_f) \subset Y$, define the map $\psi^{\#} : R_f = \mathcal{O}_Y(U) \to (\psi_{*}\mathcal{O}_X)(U)$ to be the composite

$$R_f \to \mathcal{O}_X(X)_f \to \mathcal{O}_X(\psi^{-1}U)$$

obtained by localizing φ. By Proposition I-9ii, this is enough to define a map of sheaves. Localizing further, we see that if $\psi(p) = q$, then $\psi^{\#}$ defines a local map of local rings $R_q \to \mathcal{O}_{X,p}$, and thus $(\psi, \psi^{\#})$ is a morphism of schemes. Clearly, the induced map

$$\psi^{\#}(Y) = \varphi$$

so the construction is indeed the inverse of the given one.   □

Of course this result says in particular that all the information in the category of affine schemes is already in the category of commutative rings.

COROLLARY I-37   The category of affine schemes is equivalent to the category of commutative rings with identity, with arrows reversed, the so-called *opposite category*.

EXERCISE I-38 ____

a. Using this, show that there exists one and only one map from any scheme to $\mathrm{Spec}(\mathbf{Z})$. In the language of categories, this says that $\mathrm{Spec}(\mathbf{Z})$ is the "terminal object" of the category of schemes.
b. Show that the one-point set is the terminal object of the category of sets.

For example, each point $[\mathfrak{p}]$ of $X = \operatorname{Spec} R$ corresponds to a scheme $\operatorname{Spec} \kappa(\mathfrak{p})$ that has a natural map to X defined by the composite map of rings

$$R \to R_{\mathfrak{p}} \to R_{\mathfrak{p}}/\mathfrak{p}_{\mathfrak{p}} = \kappa(\mathfrak{p})$$

Of course, the inclusion makes $[\mathfrak{p}]$ a closed subscheme iff $\mathfrak{p}$ is a maximal ideal of R (in general, $[\mathfrak{p}]$ is an infinite intersection of open subschemes of a closed subscheme).

If $\psi : Y \to X$ is a morphism of affine schemes, $X = \operatorname{Spec} R$ and $Y = \operatorname{Spec} T$, and X' is a closed subscheme of X, defined by an ideal I in R, then we define the *preimage* (sometimes, for emphasis, the "scheme-theoretic preimage") $\psi^{-1}X'$ of ψ over X' to be the closed subscheme of Y defined by the ideal $\varphi(I)T$ in T. If X' is a closed point p of X, then we call $\psi^{-1}p$ the *fiber* over X'. (We will soon see how to define fibers over arbitrary points.) The reduced subscheme of the preimage is just the set-theoretic preimage, while the scheme structure of the preimage gives a subtle and useful notion of the "correct multiplicity" with which to count the points in the preimage. The simplest classical example is given later in Exercise II-2; here we give two others.

EXERCISE I-39 ____

a. Let $\varphi : X \to Y$ be the map of affine schemes pictured below.

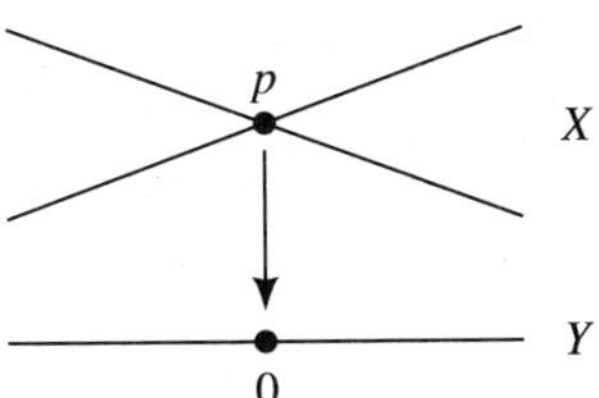

That is, $X = \operatorname{Spec}(k[x, u]/((x) \cap (u)))$ is the union of two lines meeting in a point $p = (x, u)$, while $Y = \operatorname{Spec}(k[t])$ is a line, and the map is an isomorphism on each of the lines of X; for example, it might be given by the map of rings

$$k[t] \to k[x, u]; \qquad t \mapsto x + u$$

Show that the fiber over the point $q_a = (t - a)$ is the scheme $\operatorname{Spec}(k \times k)$ consisting of two distinct points if $a \neq 0$, while the fiber over q_0—that is, the fiber containing the "double point" p—is isomorphic to $\operatorname{Spec} k[x]/(x^2)$. The fact that this is a two-dimensional algebra reflects the structure of the map locally at p.

b. Let $\varphi : X \to Y$ be the map of affine schemes pictured below.

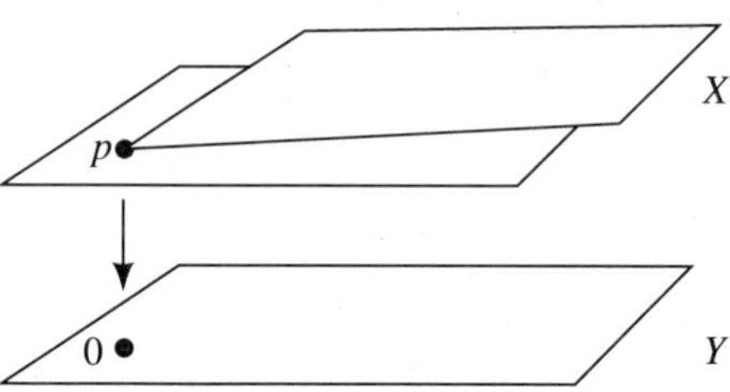

That is, $X = \mathrm{Spec}(k[x, y, u, v]/((x, y) \cap (u, v)))$ is the union of two planes in four-space meeting in a single point $p = (x, y, u, v)$, while $Y = \mathrm{Spec}(k[s, t])$ is a plane, and the map is an isomorphism on each of the planes of X; for example, it might be given by the map of rings

$$k[s, t] \to k[x, y, u, v]; \qquad t \mapsto x + u, \qquad s \mapsto y + v$$

Show that the fiber over the point $q_{a,b} = (s - a, t - b)$ is the scheme $\mathrm{Spec}(k \times k)$ consisting of two distinct points if a or $b \neq 0$, while the fiber over $q_{0,0}$—that is, the fiber containing the "double point" p—is isomorphic to $\mathrm{Spec}\, k[x, y]/(x^2, xy, y^2)$. The fact that this algebra is a three-dimensional vector space over k instead of a two-dimensional vector space (as one might expect by analogy with the previous example) reflects a deep fact about the variety X (that it is not "locally Cohen-Macaulay"). This example will be taken up again, from the point of view of flatness, in Exercise II-21.

a The Gluing Construction

Using the notion of morphism, we can construct more complicated schemes (for example, nonaffine schemes) by identifying simpler schemes along open subsets. This is a basic operation, called the *gluing construction*; it goes as follows.

Suppose that $\{X_\alpha\}_I$ is a collection of schemes, and $X_{\alpha\beta}$ is an open set in X_α for each $\beta \neq \alpha$ in I. Suppose also that we are given a family of isomorphisms of schemes

$$\psi_{\alpha\beta} : X_{\alpha\beta} \to X_{\beta\alpha} \qquad \text{for each } \alpha \neq \beta \text{ in } I$$

satisfying the *compatibility* condition

$$\psi_{\beta\gamma}\psi_{\alpha\beta} = \psi_{\alpha\gamma}$$

wherever both sides are defined. Under these circumstances we may define a scheme X by gluing the X_α along the $\psi_{\alpha\beta}$ in an obvious way—that is to say, there exists a (unique) scheme X with open subschemes isomorphic to the X_α such that the identity map on the intersections $X_\alpha \cap X_\beta \subset X$ correspond to the isomorphisms $\psi_{\alpha\beta}$.

This construction can be used, for example, to define projective schemes out of affine ones. (Another use is in the theory of toric varieties; see, for example, Kempf et al. [1973].)

In these and indeed in almost all applications, we don't really need to give the maps $\psi_{\alpha\beta}$ explicitly: we are actually given a topological space $|X|$ and a family of open subsets $|X_\alpha|$, each endowed with the structure of an affine scheme—that is, with a structure sheaf $\mathscr{O}_{X_\alpha}$ in such a way that $\mathscr{O}_{X_\alpha}(X_\alpha \cap X_\beta)$ is *naturally* identified with $\mathscr{O}_{X_\beta}(X_\alpha \cap X_\beta)$. For example, they might both be given as subsets of a fixed set. Under these circumstances it is immediate that the conditions of Corollary I-11 are satisfied, so that there is a uniquely defined sheaf $\mathscr{O}_X$ on X extending all the $\mathscr{O}_{X_\alpha}$. The pair $(|X|, \mathscr{O}_X)$ is then a scheme.

EXERCISE I-40

Show that the result of gluing schemes together as above really is a scheme.

EXERCISE I-41

Let $Y = \operatorname{Spec} k[s]$ and let $Z = \operatorname{Spec} k[t]$. Let $U \subset Y$ be the open set Y_s and let $V \subset Z$ be the open set Z_t. Let $\psi : U \to V$ be the isomorphism corresponding to the map $\mathscr{O}_Y(U) = k[s, s^{-1}] \to \mathscr{O}_Z(V) = k[t, t^{-1}]$ sending s to t, and let γ be the map sending s to t^{-1}. Let X_1 the scheme obtained by gluing together Y and Z along ψ, and let X_2 be the scheme obtained by gluing along γ instead.

Show that X_1 is not isomorphic to X_2. In fact, X_2 is the scheme corresponding to the variety $\mathbf{P}^1$, while X_1 is the affine space with a doubled origin, sometimes pictured as follows:

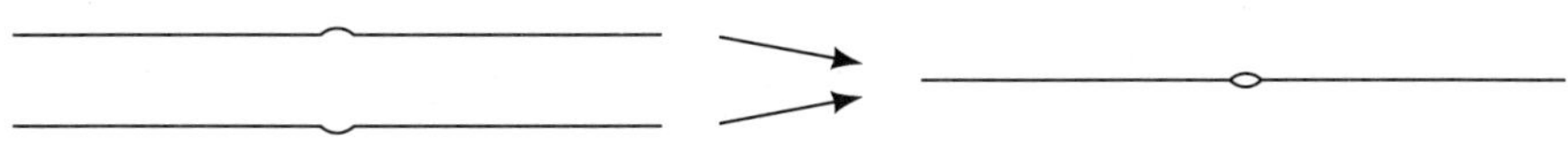

C   Relative Schemes

i Fibered Products

There is an extremely important generalization of the idea of preimage of a set under a function in the notion of the fibered product of schemes. To prepare for the definition, we first recall the situation in the category of sets.

The fibered product of two sets X and Y over a third set S—that is, of a diagram of maps of sets

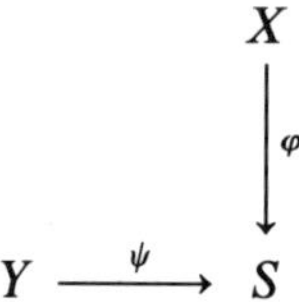

is by definition the set

$$X \times_S Y = \{(x, y) \in X \times Y \mid \varphi x = \psi y\}$$

The fibered product is sometimes called the *pullback* of X (or of $X \to S$) to Y. This construction generalizes several more elementary ones in a very useful way:

If S is a point, it gives the usual direct product.

If X, Y are both subsets of S, it gives the intersection.

If Y is a subset of S, it gives the preimage of Y in X.

If $X = Y$, it gives the set on which the maps φ, ψ are equal, the *equalizer* of the maps.

EXERCISE I-42 ____

Check these assertions!

Note that $X \times_S Y$ comes with natural projection maps to X and Y making the diagram

$$
\begin{array}{ccc}
X \times_S Y & \longrightarrow & X \\
\downarrow & & \downarrow{\scriptstyle\varphi} \\
Y & \xrightarrow{\ \psi\ } & S
\end{array}
$$

commute. Indeed, the set $X \times_S Y$ may be defined by the following universal property: among all sets Z with given maps to X and Y making the diagram

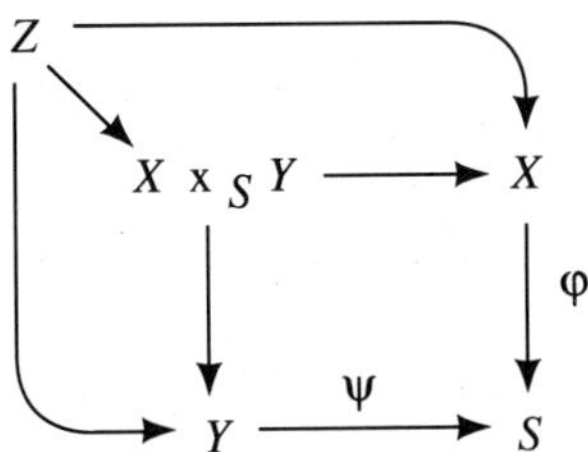

commute, $X \times_S Y$ with its projection maps is the unique "most efficient" choice in the sense that given the diagram with Z above, there is a unique map $Z \to X \times_S Y$ making the diagram

commute.

In the category of schemes we simply *define* the fibered product to be a scheme with this universal property—the universal property guarantees in particular that such a thing, with its projections to X and Y, will be unique. We can then define products, intersections, preimages, and equalizers in terms of the fibered product! However, this begs the question of whether any such object as the fibered product exists in the category of schemes. It does, and we will now describe the construction.

First, we treat the affine case. Since the category of affine schemes is, by Corollary I-37, opposite to the category of commutative rings, we must define the fibered product $X \times_S Y$ of

$$X = \operatorname{Spec} A$$

and

$$Y = \operatorname{Spec} B$$

over

$$S = \operatorname{Spec} R$$

where X and Y map to S (so that A and B are R-algebras), to be

$$X \times_S Y = \operatorname{Spec} A \otimes_R B$$

This is because the natural diagram

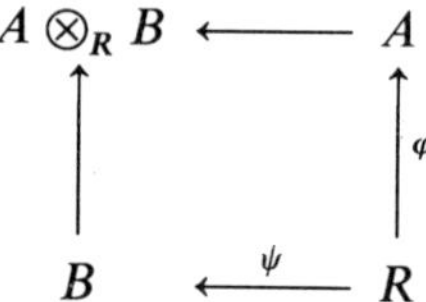

has, in an obvious sense, the opposite universal property to the one desired for the fibered product. (In fancy language, the tensor product is a fibered coproduct, or fibered sum, in the category of commutative rings.)

To check that this definition is reasonable, one may note that in the situation where Y is a closed subscheme of S defined by an ideal I, so that $B = R/I$, we have $A \otimes_R B = A/IA$, so that $X \times_S Y = \operatorname{Spec} A/IA$ is the same as the preimage of Y in X, as previously defined.

EXERCISE I-43

a. Let m, n be integers. Compute the fibered product

$$\operatorname{Spec}(\mathbf{Z}/(m)) \times_{\operatorname{Spec}(\mathbf{Z})} \operatorname{Spec}(\mathbf{Z}/(n))$$

b. Compute the fibered product $\operatorname{Spec}(\mathbf{C}) \times_{\operatorname{Spec}(\mathbf{R})} \operatorname{Spec}(\mathbf{C})$.

c. Show that for any polynomial rings $R[x]$ and $R[y]$ over a ring R, we have

$$\operatorname{Spec}(R[x]) \times_{\operatorname{Spec}(R)} \operatorname{Spec}(R[y]) = \operatorname{Spec}(R[x, y])$$

Note that in example a the underlying set of the fibered product is the fibered product of the underlying sets, but this is *not* true in examples b and c.

Now in the general case, we cover S by affine schemes $\operatorname{Spec} S_a$, and cover their preimages in X and Y by affine schemes $\operatorname{Spec} R_{ab}$ and $\operatorname{Spec} T_{ac}$, respectively, so that in a suitable sense the diagram

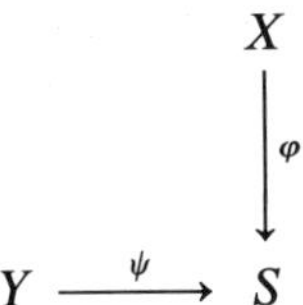

is covered by diagrams of the form

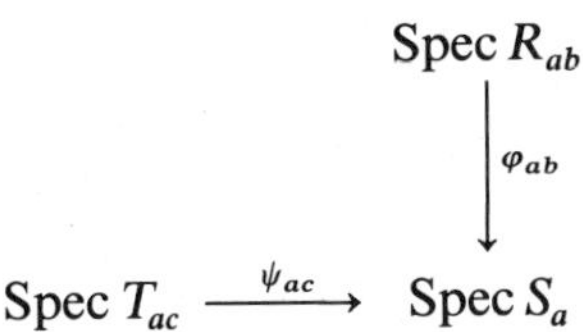

Of course, we already know that the fiber product of this last diagram is $\mathrm{Spec}(R_{ab} \otimes_{S_a} T_{ac})$. Using the idea of gluing explained at the end of the preceding section, it is easy but tedious to check that these schemes agree on overlaps and patch together to form the scheme $X \times_S Y$ as required; we omit the computation. A different approach will be sketched in Section IV-B-i.

We can use the fibered product to define the fiber of a morphism $\psi : Y \to X$ over an arbitrary point of arbitrary schemes: if p is a point of X corresponding to a prime ideal $\mathfrak{p}$ of R, then the fiber of ψ over p is the fibered product of Y and the one-point scheme $\mathrm{Spec}\,\kappa(p)$. In the case where X and Y are affine—say, $Y = \mathrm{Spec}\,T$, $X = \mathrm{Spec}\,R$—we get

$$\psi^{-1}(p) = \mathrm{Spec}\,\kappa(p) \times_X Y = \mathrm{Spec}(R_{\mathfrak{p}}/\mathfrak{p}_{\mathfrak{p}} \otimes_R T) = \mathrm{Spec}(R_{\mathfrak{p}}/\mathfrak{p}_{\mathfrak{p}} \otimes_R T/\mathfrak{p}T)$$

as a point set; this is the set of primes of T whose preimages in R are equal to $\mathfrak{p}$. More generally, we can describe the inverse image of a closed subscheme X' of X under ψ as the fibered product $X' \times_X Y$.

Another typical use of the fibered product is in studying the behavior of varieties under extension of a base field (one usually speaks in this context of a "base change" rather than a fibered product). In this setting, of which we will see a small sample in the examples in the following chapter, the notion is responsible for the great flexibility and convenience of the theory of schemes in handling arithmetic questions.

One should be aware that, as in examples b and c of Exercise I-43, the set of points of the fibered product of schemes $X \times_S Y$ is usually *not* equal to the fibered product (in the category of sets) of the sets of points of X and Y. This is no terrible pathology but simply reflects the fact that the theory of functions $f(x, y)$ of two variables is much richer than the theory of functions of the form $g(x)h(y)$. In any case, the definitions of Chapter IV provide a viewpoint from which this oddity disappears.

ii The Category of S-Schemes

Just as in the case of sets, we can use the fibered product to define an absolute product by taking S to be a terminal object in the category of schemes—that is, a scheme such that every scheme has a unique map to S. By Exercise I-38 the terminal object in the category of schemes is Spec $\mathbf{Z}$. However, the absolute product has some rather surprising properties. We have already seen in Exercise I-43a cases (when m and n are relatively prime) where the product in this sense of nonempty sets may be empty! There are other peculiarities as well: for example, the dimension of an irreducible scheme can be defined as the Krull dimension of the coordinate ring of any of its affine open sets. One might expect the product

$$X \times Y = X \times_{\text{Spec } \mathbf{Z}} Y$$

of two schemes to have dimension equal to the sum of the dimensions of X and Y. But in fact we have the result in the next exercise.

EXERCISE I-44* ____

Show that if $X = \operatorname{Spec} R$ and $Y = \operatorname{Spec} S$, where R and S are domains finitely generated as $\mathbf{Z}$-algebras and containing $\mathbf{Z}$, then

$$\dim X \times Y = \dim X + \dim Y - \dim \operatorname{Spec} \mathbf{Z}$$

$$= \dim X + \dim Y - 1$$

This oddity and many like it can be eliminated by a rather trivial but convenient generalization of our definitions: we often wish to work with schemes X "defined over a given field (or ring) k," or "k-schemes." Of course, we will then use only morphisms that respect this structure. Informally, this just means that we consider X together with a k-algebra structure on $\mathcal{O}_X(X)$ and morphisms respecting these structures.

In this category, $\operatorname{Spec} k$ is the terminal object and the absolute product is the fibered product over $\operatorname{Spec} k$. If k is a field, then the product in the

II

Examples

A Reduced Schemes Over Algebraically Closed Fields

We will start our series of examples with the one that the concept of scheme is intended to generalize: the classical notion of an affine variety over an algebraically closed field k. In our present context, this means considering schemes of the form $\operatorname{Spec} R$, where R is the coordinate ring of a variety X—that is, a finitely generated, reduced algebra over k. This scheme is sometimes called the *scheme associated to the variety* X; such schemes are sometimes referred to just as varieties. After describing these, we will go on in succeeding sections to consider the ways in which schemes may differ from this basic model.

The k-scheme associated to an affine variety over an algebraically closed field k is an equivalent object to the variety; either one determines and is determined by its coordinate algebra, which is the same for both. However, even in this case, classical notions such as the intersection of varieties and the fibers of maps are given a more precise meaning in the theory of schemes. We will see examples of this phenomenon in this and succeeding sections.

i Affine Spaces

We start with the scheme $\mathbf{A}_k^n := \operatorname{Spec} k[x_1, \ldots, x_n]$ with k an algebraically closed field. This scheme is called *affine n-space* over k. Where the field k is indifferent, or is clear from context, we usually write simply $\mathbf{A}^n$.

We will make use of a standard but nontrivial result from algebra, a form of Hilbert's "Nullstellensatz" (Matsumura, 1986, Theorem 5.3).

THEOREM II-1 (Nullstellensatz.)   Let k be any field. If $\mathfrak{m}$ is a maximal ideal of a polynomial ring $k[x_1, \ldots, x_n]$ (or, geometrically, p is a closed point of any subvariety of an affine space over a field k), then $k[x_1,\ldots,x_n]/\mathfrak{m} = \kappa(p)$ is a finite-dimensional vector space over k.

In our case, with k algebraically closed, this implies that $\kappa(p) = k$. Thus writing λ_i for the image of x_i in $\kappa(p)$, we see that $\mathfrak{m} = (x_1 - \lambda_1,\ldots,x_n - \lambda_n)$. In this way the closed points of $\mathbf{A}_k^n$ correspond to n-tuples of elements of k, as one should expect. We will sometimes refer to "the point $(\lambda_1,\ldots,\lambda_n)$" instead of "the point $[(x_1 - \lambda_1,\ldots,x_n - \lambda_n)]$."

To begin with dimension 1, the affine line

$$\mathbf{A}_k^1 = \operatorname{Spec} k[x]$$

looks almost exactly like its classical counterpart, the algebraic variety also called the affine line. It contains one closed point for each value $\lambda \in k$. The Zariski topology on the set of closed points is the same as the classical Zariski topology on the variety: the open sets are the complements of finite sets. The scheme $\mathbf{A}^1$ differs from the variety only in that the scheme contains one more point, the *generic* point of $\mathbf{A}^1$, corresponding to the ideal (0).

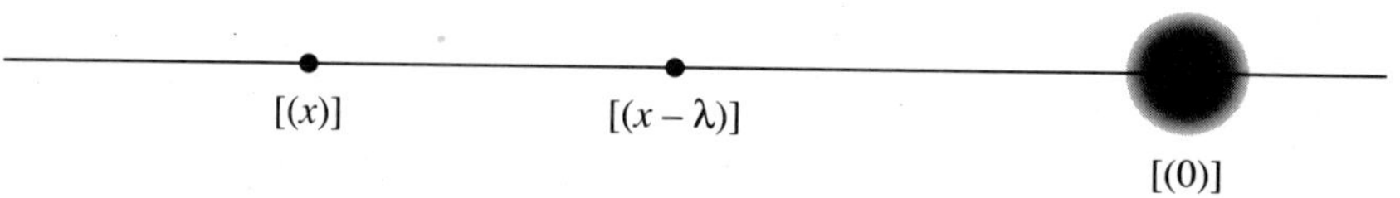

Note that the closure of the point $[(0)]$ is all of $\mathbf{A}^1$, so that the closed subsets of $\mathbf{A}^1$ are exactly the finite subsets of $\mathbf{A}^1$-$\{[(0)]\}$.

The affine plane $\mathbf{A}_k^2 = \operatorname{Spec} k[x,y]$ is also similar to its counterpart variety, but now the additional points of the scheme are more numerous and behave in more interesting ways. We have as before closed points, coming from the maximal ideals $(x - \lambda, y - \mu)$, which correspond to the points (λ, μ) in the ordinary plane. There are now, however, two types of nonclosed points. To begin with, for each irreducible polynomial $f(x,y) \in k[x,y]$ we have a point corresponding to the prime ideal $(f) \subset k[x,y]$, whose closure consists of the point itself and all the closed points (λ, μ) with $f(\lambda, \mu) = 0$. The point $[(f)]$ is called the *generic* point of this set; more generally, any point in a scheme is called the generic point of its closure. As compared to the variety $\mathbf{A}^2$, we have added one more point for every irreducible plane curve C. This new point lies in the closure of (the set of closed points on) that curve, and its closure contains this set of closed points. Finally, we have as before a point corresponding to the zero ideal, the generic point of $\mathbf{A}^2$, whose closure is all of $\mathbf{A}^2$.

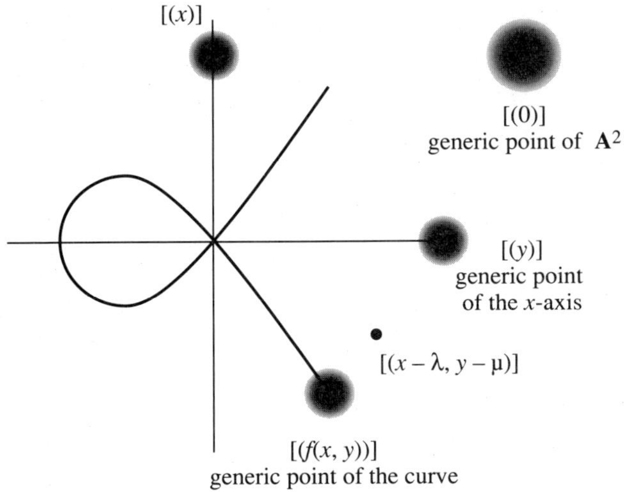

Since $k[x, y] = k[x] \otimes_k k[y]$, we have by definition

$$\mathbf{A}_k^2 = \mathbf{A}_k^1 \times_{\mathrm{Spec}\, k} \mathbf{A}_k^1$$

Note that even here, though it is certainly clear that the fibered product is the correct notion of product, the set of points of the fibered product is not the fibered product of the sets of points of the factors.

The situation with the affine spaces $\mathbf{A}_k^n = \mathrm{Spec}\, k[x_1, \ldots, x_n]$ is a straightforward extension of the last case: geometrically, we can see the scheme $\mathbf{A}^n$ as the classical affine n-space, with one point p_Σ added for every positive-dimensional irreducible subvariety Σ of n-space. As above, p_Σ will lie in the closure of the locus of closed points in Σ and contain all these points, as well as the generic points of the subvarieties of Σ, in its closure.

More generally, suppose $X \subset \mathbf{A}^n$ is any affine variety, with ideal $I \subset k[x_1, \ldots, x_n]$ and coordinate ring $R = k[x_1, \ldots, x_n]/I$. We can associate to X the affine scheme $\mathrm{Spec}\, R$; the quotient map $k[x_1, \ldots, x_n] \to R$ expresses this as a subscheme of $\mathbf{A}^n$. This scheme is, as in the case of $\mathbf{A}^n$ itself, just like the variety X except that we have added one new generic point p_Σ for every positive-dimensional irreducible subvariety $\Sigma \subset X$.

Fibers, and more generally preimages, are among the most common ways that schemes other than varieties may arise even in the context of classical geometry.

EXERCISE II-2

Consider the map of the affine line $\operatorname{Spec} k[x]$ to itself given by $x \to x^2$. Show that the scheme-theoretic fiber over the point 0 is the subscheme of the line defined by the ideal (x^2).

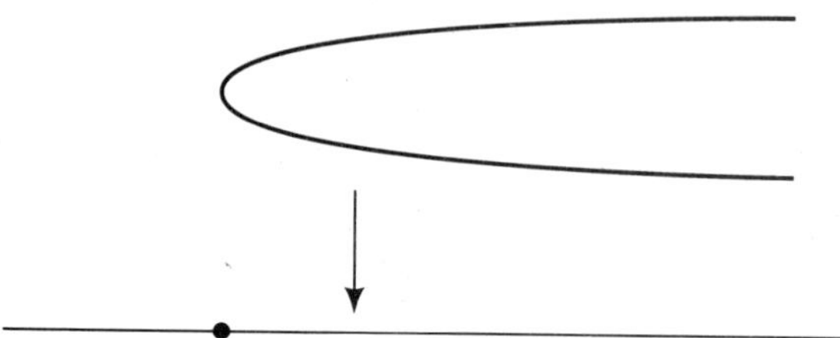

Among all schemes, those associated to affine varieties over algebraically closed fields may be characterized as spectra of rings R that are

a. finitely generated
b. reduced algebras
c. over a field
d. that is algebraically closed.

To get a sense of what more general schemes look like, and what they are good for, we will in the remainder of this section and the next consider what may happen if we remove these four restrictions. We will consider primarily examples in which exactly one of the hypotheses fails, since for the most part an understanding of these basic cases will enable one to understand the general case; we will occasionally mention more complex examples in exercises.

ii Local Schemes

Our first collection of examples of schemes other than varieties is provided by the spectra of local rings, called *local schemes*. The examples we will consider here are spectra of rings that are reduced algebras over an algebraically closed field but not, in general, finitely generated. Local schemes are for the most part technical tools in the study of other, more geometric schemes; they are often used to focus attention on the local structure of an affine scheme. The extra points we have added to classical varieties show up even more strikingly in the following examples, where in each case there is only one closed point. It would, of course, be a mistake to try to picture these schemes as geometric objects with just one point. Rather, they should be seen as germs of varieties. The phenomenon of having only one closed point is not some novelty invented by algebraists but is already present if one considers such a familiar object as the germ of a point x on a complex analytic manifold; here one pictures a "sufficiently small" neighborhood of x, in which, for example, each curve through x can be plainly distinguished, even though no other definite points beside x belong to every neighborhood. We will see that the same kind of picture is valid for the spectrum of a local ring.

Consider first the localization $k[x]_{(x)}$ of the ring $k[x]$ at the maximal ideal (x), and let $X = \operatorname{Spec} k[x]_{(x)}$. The space $|X|$ has only two points: the closed point corresponding to the maximal ideal (x), and the open point corresponding to (0), which contains the point $[(x)]$ in its closure. The inclusion of $k[x]$ in $k[x]_{(x)}$ induces a map $X \to \mathbf{A}^1$, so that we may think of X as a subscheme of $\mathbf{A}^1$ (though $|X|$ is neither open nor closed in $|\mathbf{A}^1|$.) The subscheme X is "local" in that it is the intersection of all the open subsets of $\mathbf{A}^1$ containing the point (x); so that, for example, the regular functions on X are exactly the rational functions on $\mathbf{A}^1$ regular at the point (x)—that is, they are the elements of $\mathcal{O}_{\mathbf{A}^1}(U)$ for some neighborhood U of the point $0 = [(x)]$ in $\mathbf{A}^1$. In these senses, X is the germ of $\mathbf{A}^1$ at the origin.

Next, consider the scheme $X = \operatorname{Spec} R$, where $R = k[x, y]_{(x,y)}$ is the localization of $k[x, y]$ at the maximal ideal corresponding to the point $(0, 0)$. As in the previous example, we have a map $X \to \mathbf{A}^2$, in terms of which we can think of X as the intersection of all open subschemes of $\mathbf{A}^2$ containing the closed point $(0, 0)$. Again, X has only one closed point; but unlike the preceding example, X has infinitely many nonclosed points: we have one such point for every irreducible curve C in the plane passing through the origin. Subschemes of X are thus germs at $(0, 0)$ of subschemes of $\mathbf{A}^2$ and X itself is the germ of $\mathbf{A}^2$ at the origin.

There are analogous constructions in $\mathbf{A}^n$, and more generally for any subscheme of $\mathbf{A}^n$: if $X = \operatorname{Spec} k[x_1, \ldots, x_n]/I \subset \mathbf{A}^n$ is the scheme associated to the affine variety with ideal $I \subset k[x_1, \ldots, x_n]$ and $\mathfrak{m} = (x_1 - a_1, \ldots, x_n - a_n)$ a maximal ideal corresponding to a closed point of X, we can consider the scheme $\operatorname{Spec} k[x_1, \ldots, x_n]_{\mathfrak{m}}/I_{\mathfrak{m}}$ as a germ of a neighborhood of $[\mathfrak{m}]$ in X. The point is that while we can talk about germs of functions on a space at a point in many contexts, in scheme theory it is the case that the germ is again a scheme in its own right.

For some purposes, the local schemes introduced in this way are not local enough; the local ring of a scheme at a point still contains a lot of information about the global structure of the scheme. For example, the germs of a nonsingular variety X at various closed points will not in general be isomorphic schemes,[1] although if X^{an} denotes the complex analytic variety defined by the same equations (or indeed any analytic manifold), then the

---

[1] This has nothing to do with schemes but is already the case for varieties over $\mathbf{C}$: for example, it is so already for the general plane curve of degree $d \geq 4$.

germs of X^{an} at any two points are isomorphic. This phenomenon occurs essentially because the open sets used to define the germs of X are so big. To get a more local picture within the setting of schemes, we can look at the schemes associated to power series rings: for example, instead of looking at the germ $X = \operatorname{Spec} k[x, y]_{(x, y)}$ of a neighborhood of the origin in $\mathbf{A}^2$ above, we can consider the scheme $Y = \operatorname{Spec} k[[x, y]]$. As in the previous case, this scheme has one closed point $[(x, y)]$ and one generic point $[(0)]$ whose closure is all of Y; in addition, it has one point for every irreducible power series $\sum a_{i,j} x^i y^j$ in x and y. The maps

$$k[x, y] \hookrightarrow k[x, y]_{(x, y)} \hookrightarrow k[[x, y]]$$

give maps $Y \to X \to \mathbf{A}^2$; we think of the Y as a "smaller" neighborhood of the origin than X. (Note, however, that X and Y are neither closed subschemes nor open subschemes of $\mathbf{A}^2$.) For example, while the curve corresponding to the prime ideal $(y^2 - x^3 - x^2)$ is irreducible in X, because the curve in $\mathbf{A}^2$ defined by this equation is, the preimage in Y of this curve *is* the (nontrivial) union of two curves in Y (as long as the characteristic of k is not 2):

Spec $k[x, y]_{(x,y)}/(y^2 - x^3 - x^2) \subset$ Spec $k[x, y]_{(x,y)}$
is irreducible;

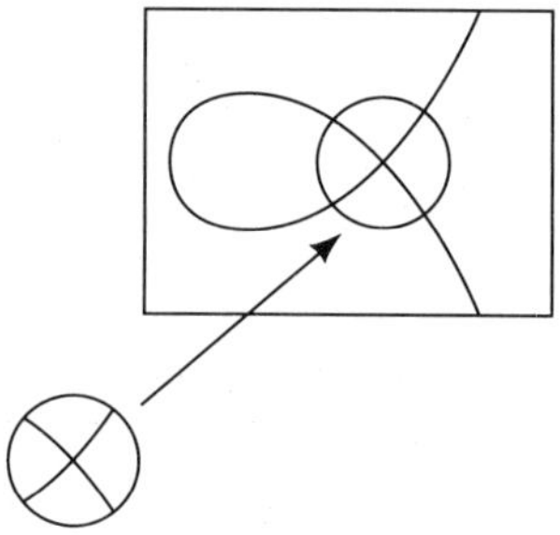

its preimage
Spec $k[[x, y]]/(y^2 - x^3 - x^2) \subset$ Spec $k[[x, y]]$
is not.

This is because $x^2 + x^3$ has the square root

$$u = x + \frac{x^2}{2} - \frac{x^3}{8} + \cdots$$

in the power series ring. Thus we can factorize $y^2 - x^3 - x^2$ as

$$y^2 - x^3 - x^2 = (y - u)(y + u)$$

so the scheme $\operatorname{Spec} k[[x, y]]/(y^2 - x^3 - x^2)$ is reducible.

Of course, Y must have "more" curves than X for such things to be possible. The following exercise amplifies on this fact.

EXERCISE II-3 ____

a. With $u = \sqrt{x^2 + x^3}$ as above, what is the image of $[(y - u)]$ in $\operatorname{Spec} k[x, y]$?

b. In general, verify that under the map $Y \to X$ above, the inverse image of a point corresponding to an irreducible curve $C \subset \mathbf{A}_k^2$ consists of the set of analytic branches of C at the origin (see Walker [1978] or Brieskorn and Knörrer [1986] for a discussion of the notion of branches of a curve).

c. Show that the image of the point $(y - \sum x^n/n!)$ of Y is the generic point of $\mathbf{A}^2$.

Here is yet another important example of a local scheme. One problem with the scheme Y above is that the points described in part b of Exercise II-3 are extraneous from an algebraic point of view. To avoid this, we may work with the spectrum Z of the ring $H \subset k[[x, y]]$ of power series that satisfies some algebraic equation over $k(x, y)$, the field of rational functions. Called the *Henselization* of X, Z sits in between Y and X in the sense that we have a series of maps

$$Y \to Z \to X \to \mathbf{A}^2$$

The usefulness of this construction (which we will not pursue further) is that H is the union of algebras finitely generated over k, so that Z is the inverse limit of schemes coming from ordinary varieties. See Artin (1971) for further information.

EXERCISE II-4 ____

In the case $k = \mathbf{C}$, how does the spectrum of the ring of convergent power series fit into this picture?

B Reduced Schemes Over Non–Algebraically Closed Fields

We now consider what happens when we look at the spectrum of a finitely generated, reduced algebra over a field k that is not algebraically closed. The interest in these came originally from number theory, and, of course, it predates scheme theory very substantially! For example, the study of rational quadratic forms, an old subject in number theory, can be thought of as the study of varieties over the rational numbers defined by a quadratic equation. Cubic forms in three variables over the rationals still make up a very active number-theoretic research topic, now mostly pursued through the theory of elliptic curves over $\mathbf{Q}$. The basic objects themselves are varieties over $\mathbf{Q}$ (or schemes over $\mathbf{Z}$, a situation we'll return to later), but in the course of handling them, number theorists frequently make use of all the base rings shown in the following diagram, along with many intermediate fields and rings:

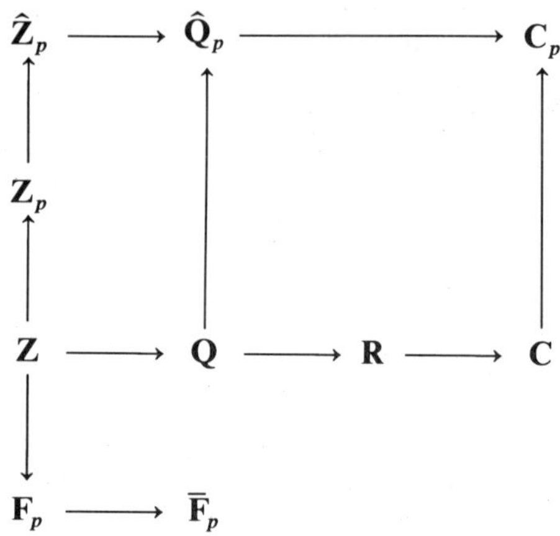

The theory of schemes provides a particularly flexible and convenient framework for handling these many "changes of base." Also, a nice variety reduced mod p may suddenly become something nonreduced—something that requires the theory of schemes more fully.

To start with the simplest case, consider $\mathbf{A}_{\mathbf{R}}^1 = \operatorname{Spec} \mathbf{R}[x]$. Using the Nullstellensatz, we see that there are two kinds of maximal ideals in $\mathbf{R}[x]$: those whose residue class field is $\mathbf{R}$, which have the form $(x - \lambda)$ for $\lambda \in \mathbf{R}$, and those whose residue class field is $\mathbf{C}$, which have the form $(x^2 + \mu x + v)$, for μ and $v \in \mathbf{R}$ with $\mu^2 - 4v < 0$. The latter type of ideals may also be written in the form $((x - z)(x - \bar{z}))$, for $z \in \mathbf{C}$ not real. The closed points of $\mathbf{A}^1$ thus correspond either to real numbers or to conjugate pairs of nonreal complex numbers. Finally, $\mathbf{A}^1$ has again a unique nonclosed point corresponding to the prime (0), whose closure is all of $\mathbf{A}^1$.

Next, we turn to the affine plane over $\mathbf{R}$, $\mathbf{A}_{\mathbf{R}}^2 = \operatorname{Spec} \mathbf{R}[x, y]$, and consider a closed point given by a maximal ideal $\mathfrak{m}$ of $\mathbf{R}[x, y]$. Again by the Nullstellensatz, the residue class field of $\mathfrak{m}$ is either $\mathbf{R}$ or $\mathbf{C}$, and the composite map

$$\mathbf{R} \to \mathbf{R}[x, y]/\mathfrak{m} \cong \begin{cases} \mathbf{R} \\ \text{or} \\ \mathbf{C} \end{cases}$$

is either the identity or the inclusion of **R** in **C**. Taking λ and μ to be the images of x and y, we see that in the former case $\mathfrak{m}$ corresponds to the ordinary point (λ, μ) in $\mathbf{R}^2$. But in the latter case $\mathfrak{m}$ "corresponds to both (λ, μ) and $(\bar{\lambda}, \bar{\mu})$"; put differently, the map $\mathbf{R}[x, y] \to \mathbf{C}$ sending x, y to λ, μ has the same kernel as the one sending x, y to $\bar{\lambda}$, $\bar{\mu}$, since they differ by the automorphism of **C** over **R**.

It is not difficult to give generators for the maximal ideals described above. If $\mathbf{R}[x, y]/\mathfrak{m} \cong \mathbf{R}$, then clearly $\mathfrak{m} = (x - \lambda, y - \mu)$. In the other case, suppose first that λ is real. Then μ must satisfy an irreducible real quadratic polynomial $y^2 + ay + b$, so $\mathfrak{m}$ contains the ideal $(x - \lambda, y^2 + ay + b)$. But this last ideal is immediately seen to be prime (for example, by factoring out $x - \lambda$ first), so $\mathfrak{m} = (x - \lambda, y^2 + ay + b)$. Of course, a similar result holds if the image of y is real.

Finally, suppose that μ and λ are both nonreal. Of course, $\mathfrak{m}$ contains the irreducible polynomials $f(x)$ and $g(y)$ satisfied by μ and λ, but since $g(y)$ factors as

$$g(y) = (y - \mu)(y - \bar{\mu})$$

in $\mathbf{R}[x]/(f(x)) \cong \mathbf{C}$, the ideal $(f(x), g(y))$ is not prime! The picture here, over the complex numbers, is as follows:

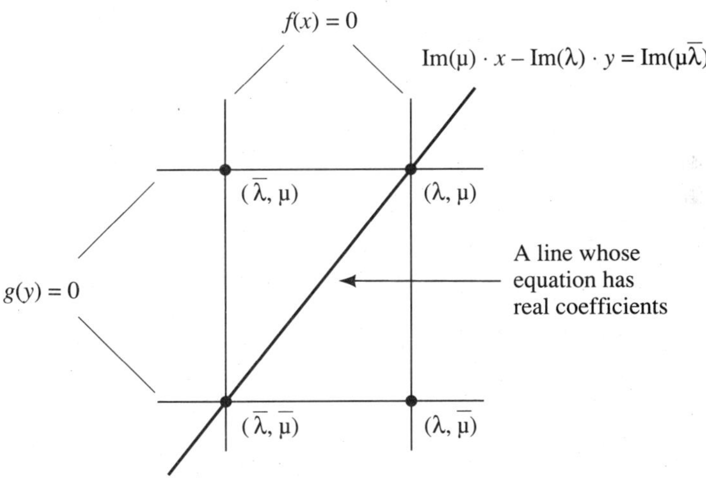

The loci defined by $f(x) = 0$ and $g(y) = 0$ are unions of two vertical and two horizontal lines, respectively, and intersect in the four points (λ, μ), $(\bar{\lambda}, \mu)$, $(\lambda, \bar{\mu})$, and $(\bar{\lambda}, \bar{\mu})$. But the polynomial

$$h(x, y) = \mathrm{Im}(\mu) \cdot x - \mathrm{Im}(\lambda) \cdot y - \mathrm{Im}(\mu\bar{\lambda})$$

defining the line joining the two points (λ, μ) and $(\bar{\lambda}, \bar{\mu})$ has real coefficients.

The ideal

$$(f(x), h(x, y)) = (g(y), h(x, y)) \subset \mathbf{R}[x, y]$$

thus strictly contains the ideal $(f(x), g(y))$; and this ideal is the maximal ideal $\mathfrak{m}$ we seek, as one checks by working in

$$\mathbf{R}[x] \cong \mathbf{R}[x, y]/(h) \cong \mathbf{R}[y]$$

[for these isomorphisms, note that $(\bar{\mu} - \mu)$ and $(\bar{\lambda} - \lambda)$ are both nonzero].

In sum, then, the closed points of $\mathbf{A}_\mathbf{R}^2$ correspond either to points (λ, μ) of $\mathbf{A}_\mathbf{C}^2$ with λ and μ real, or to (unordered) pairs of points (z, w) and $(\bar{z}, \bar{w}) \in \mathbf{A}_\mathbf{C}^2$ with at least one of z, w not real. To put it another way, closed points of $\mathbf{A}_\mathbf{R}^2$ correspond to orbits of the action of complex conjugation on the points of $\mathbf{A}_\mathbf{C}^2$. (Note, in particular, that the closed points of $\mathbf{A}_\mathbf{R}^2$ are not ordered pairs of closed points of $\mathbf{A}_\mathbf{R}^1$!) Observe also that the residue field is $\mathbf{R}$ at the points of $\mathbf{A}_\mathbf{R}^2$ corresponding to points (λ, μ) with λ, μ real, while at points of $\mathbf{A}_\mathbf{R}^2$ corresponding to pairs of complex conjugate points of $\mathbf{A}_\mathbf{C}^2$ the residue field is $\mathbf{C}$.

EXERCISE II-5* ──

Show that the nonclosed points of $\mathbf{A}_\mathbf{R}^2$ are all either

a. $[(0)]$, whose closure is all of $\mathbf{A}_\mathbf{R}^2$, or
b. the point $[(f)]$ of $\mathbf{A}_\mathbf{R}^2$ corresponding to an irreducible polynomial $f \in \mathbf{R}[x, y]$.

Those of type b may or may not remain irreducible in $\mathbf{C}[x, y]$, so that a nonclosed point (f) in $\mathbf{A}_\mathbf{R}^2$ will correspond either to a single nonclosed point in $\mathbf{A}_\mathbf{C}^2$ (if f remains irreducible in $\mathbf{C}[x, y]$) or to two nonclosed points in $\mathbf{A}_\mathbf{C}^2$ (if f may be written as a product $g \cdot \bar{g}$ with $g \in \mathbf{C}[x, y]$). The closed points in the closure of such a nonclosed point may be either of both types above or only of the second. Give examples with all these possibilities.

The situation in general follows the lines of these examples: if k is any field, $\bar{k}$ its algebraic closure, and $G = \mathrm{Gal}(\bar{k}/k)$ the corresponding Galois group, then the points of $\mathbf{A}_k^n$ correspond to orbits of the action of G on the points of $\mathbf{A}_{\bar{k}}^n$ (see, for example, Nagata [1962, Theorem 10.3]). The closed points correspond to orbits of closed points, the orbits being finite. The residue field at the point P corresponding to such an orbit, moreover, is isomorphic to the fixed field of the action on $\bar{k}$ of the subgroup G_p fixing a point of that orbit. For example, the closed points of $\mathbf{A}_\mathbf{Q}^1$ correspond to algebraic numbers modulo conjugacy; and for a prime number $p \in \mathbf{Z}$ the closed points of $\mathbf{A}^1_{\mathbf{F}_p}$ correspond to the orbits of the Frobenius automorphism of the algebraic closure of $\mathbf{F}_p = \mathbf{Z}/(p)$. (This last is the set consisting of 0 and the orbits of multiplication by p on the multiplicative group $\bar{k}^*$, which may be described as the inductive limit of all cyclic groups of order prime to p or as the p-torsion free part of $\mathbf{Q}/\mathbf{Z}$.)

EXERCISE II-6

An inclusion of fields $K \hookrightarrow L$ induces a map $\mathbf{A}^n_L \to \mathbf{A}^n_K$. Find the images in $\mathbf{A}^2_{\mathbf{Q}}$ of the following points of $\mathbf{A}^2_{\overline{\mathbf{Q}}}$ under this map.

a. $(x - \sqrt{2}, y - \sqrt{2})$
b. $(x - \sqrt{2}, y - \sqrt{3})$
c. $(x - \zeta, y - \zeta^{-1})$, ζ a pth root of unity, p prime
d. $(\sqrt{2}x - \sqrt{3}y)$
e. $(\sqrt{2}x - \sqrt{3}y - 1)$

Where feasible, draw pictures.

EXERCISE II-7

We say that a subscheme $X \subset \mathbf{A}^n_K$ is *absolutely irreducible* if the inverse image of X in $\mathbf{A}^n_{\overline{K}}$ is irreducible (more generally, we say any scheme X over $\operatorname{Spec} K$ is absolutely irreducible if the fiber product $X \times_{\operatorname{Spec} K} \operatorname{Spec} \overline{K}$ is irreducible). For each of the following subschemes of $\mathbf{A}^2_{\mathbf{Q}}$, say whether it is reducible, irreducible but not absolutely irreducible, or absolutely irreducible.

a. $(x^2 - y^2)$
b. $(x^2 + y^2)$
c. $(x^2 + y^2 - 1)$
d. $(x + y, xy - 2)$
e. $(x^2 - 2y^2, x^3 + 3y^3)$

Finally, here is an example that combines the notions of local schemes and schemes over non-algebraically closed fields. Classically, a plane curve $X \subset \mathbf{A}^2_{\mathbf{C}}$ was said to have a *node* at the origin if in some analytic neighborhood of the origin the locus of complex points of X consisted of two smooth arcs intersecting transversely at $(0,0)$. In the language of schemes, this is the same as saying that the fiber product of X with the formal neighborhood $\operatorname{Spec} \mathbf{C}[[x, y]] \to \operatorname{Spec} \mathbf{C}[x, y] = \mathbf{A}^2$ is isomorphic to $\operatorname{Spec} \mathbf{C}[[u, v]]/(uv)$.

Consider now a curve in the real plane $X \subset \mathbf{A}^2_{\mathbf{R}}$. We say in this case that X has a node at the origin if the corresponding complex curve $X \times_{\operatorname{Spec} \mathbf{R}} \operatorname{Spec} \mathbf{C} \subset \mathbf{A}^2_{\mathbf{C}}$ does. In this case, the formal neighborhood

$$X_0 = X \times_{\mathrm{Spec}\ \mathbf{R}[x,y]} \mathrm{Spec}\ \mathbf{R}[[x,y]]$$

may have either one of two nonisomorphic forms: it may be isomorphic to $\mathrm{Spec}\ \mathbf{R}[[u,v]]/(uv)$ or to $\mathrm{Spec}\ \mathbf{R}[[u,v]]/(u^2 + v^2)$. The former is the case if the locus of real points of X (that is, the locus of points with residue field $\mathbf{R}$) looks in an analytic neighborhood of $(0,0)$ like two smooth real arcs intersecting transversely at $(0,0)$; classically, such a point was called a *crunode* of X. The latter is the case if the origin is isolated as a real point of X; this was called an *acnode* in the past.

EXERCISE II-8

Verify the assertions made above: specifically, show that if X is a curve in $\mathbf{A}_{\mathbf{C}}^2$ with a node at the origin, then the formal neighborhood $X \times_{\mathrm{Spec}\ \mathbf{C}[x,y]}$ $\mathrm{Spec}\ \mathbf{C}[[x,y]]$ is isomorphic to $\mathrm{Spec}\ \mathbf{C}[[u,v]]/(uv)$; and that if $X \subset \mathbf{A}_{\mathbf{R}}^2$ is a real plane curve with a node at the origin, then the formal neighborhood $X \times_{\mathrm{Spec}\ \mathbf{R}[x,y]} \mathrm{Spec}\ \mathbf{R}[[x,y]]$ has one of the two forms above. Show that there are infinitely many curves $X \subset \mathbf{A}_{\mathbf{Q}}^2$ with nodes at the origin having non-isomorphic formal neighborhoods.

C   Nonreduced Schemes

We now leave the realm of objects that could be treated in the theory of varieties to look at some examples of affine schemes $\mathrm{Spec}\ R$, where R is a finitely generated algebra over an algebraically closed field k but may have nilpotents. The phenomena here are much less familiar, and we will spend rather more effort on them.

Schemes of this type arise already in quite simple geometric contexts: for example, the multiple points treated first below occur already as intersections of two ordinary varieties and as "degenerate" fibers of maps (recall that a fiber of a map is just the preimage of a point). One of the most important applications of nonreduced schemes is to the theory of families of varieties: deformation theory and moduli theory. We will introduce the key notion of flatness below and explain how to take "flat limits" of one-parameter families of varieties. Finally, we will give some examples of nonreduced schemes that are interesting objects in themselves.

To start with the easiest cases, we will focus first on subschemes of affine space $\mathbf{A}^n$ supported at the origin—equivalently, given by ideals I whose zero locus $V(I)$ consists, as a set, just of $(0,\dots,0)$.

i Double Points

The simplest such scheme is the subscheme X of $\mathbf{A}^1$ defined by the ideal (x^2)—that is, the scheme $\operatorname{Spec} k[x]/(x^2)$, viewed as a subscheme of $\mathbf{A}^1$ via the map induced by the quotient map $k[x] \to k[x]/(x^2)$. This scheme has only one point, corresponding to the ideal (x), but it is different—both as a subscheme of $\mathbf{A}^1$ and as an abstract scheme—from the scheme $\operatorname{Spec} k[x]/(x) = \operatorname{Spec} k$. As an abstract scheme, we can see the difference in that there exist regular functions (such as x) on X that are not equal to zero but that have value 0 at the one point of X; of course, any such function will have square 0. As a subscheme of $\mathbf{A}^1$, the difference is that a function $f \in k[x]$ on $\mathbf{A}^1$ vanishes on X if and only if both f and its first derivative vanish at 0. The data of a function on X is thus the data of the values at 0 of both a function on $\mathbf{A}^1$ *and* its first derivative. Possibly for this reason, X is sometimes called the *first-order neighborhood of* 0 *in* $\mathbf{A}^1$.

More generally, for any n the ideal (x^n) defines a subscheme $X_n \subset \mathbf{A}^1$ with coordinate ring $k[x]/(x^n)$; a function $f(x)$ on $\mathbf{A}^1$ vanishes on X if and only if the value of f at 0 vanishes together with the values of the first $n-1$ derivatives of f.

The picture of double points may become clearer if we consider subschemes of $\mathbf{A}^2 = \operatorname{Spec} k[x, y]$ isomorphic to X and supported at the origin. Let $Y \subset \mathbf{A}^2$ be such a subscheme, $R = \mathcal{O}_Y(Y) \cong k[\varepsilon]/(\varepsilon^2)$ its coordinate ring, and

$$\varphi : k[x, y] \to R$$

the surjection defining the inclusion of Y in $\mathbf{A}^2$. Since the inverse image of the unique maximal ideal $\mathfrak{m}$ of R is the ideal $(x, y) \subset k[x, y]$ corresponding to the origin, and since $\mathfrak{m}^2 = 0$ in R, the map φ will factor through a map

$$\overline{\varphi} : k[x, y]/(x^2, xy, y^2) \to R$$

Equivalently, Y must be contained in the subscheme $Z = \operatorname{Spec} k[x, y]/(x^2, xy, y^2)$. On the other hand, the ring $k[x, y]/(x^2, xy, y^2)$ is a three-dimensional vector space over k, whereas R is only two-dimensional. It follows that the kernel of φ will contain a homogeneous linear form in x and y—that is, we can write

$$X = X_{\alpha, \beta} = \operatorname{Spec} k[x, y]/(x^2, xy, y^2, \alpha x + \beta y) \hookrightarrow \mathbf{A}^2$$

for some pair $[\alpha, \beta] \in k^2$.

The subscheme $X_{\alpha,\beta}$ can be characterized as follows:

i. The subscheme of $\mathbf{A}^2$ associated to the ideal of functions $f \in k[x, y]$ that vanish and have partial derivatives satisfying

$$\beta \frac{\partial f}{\partial x} - \alpha \frac{\partial f}{\partial y} = 0$$

at the origin (since this implies that $f = c \cdot (\alpha x + \beta y) +$ higher-order terms).
ii. Or the image of the subscheme $X \subset \mathbf{A}^1$ of the last example under the inclusion of $\mathbf{A}^1$ in $\mathbf{A}^2$ given by $x \mapsto (\beta x, -\alpha x)$.

Informally, we say that the subscheme $X_{\alpha,\beta}$ consists of the point $(0,0)$ and an "infinitely near point" in the direction specified by the line $(\alpha x + \beta y = 0)$; we would draw $X_{\alpha,\beta}$ as the small arrow in the rather traditional picture:

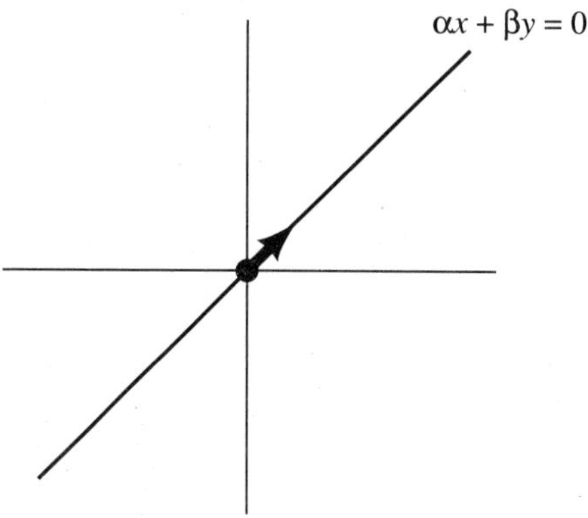

This is intended to represent a point with a distinguished one-dimensional subspace of the tangent space to the plane at that point—there is actually no distinguished tangent vector, despite the impression given by the arrow.

How do schemes such as $X_{\alpha,\beta}$ arise in practice? One way is as intersections of curves. For example, when we want to work with the intersection of a line L and a conic C that happen to be tangent, it is clearly unsatisfactory to take their intersection in the purely set-theoretic sense; a line and a conic should meet twice. Nor is it completely satisfactory to describe $C \cap L$ as their point of intersection "with multiplicity two"; the intersection should determine L. The satisfactory definition is that $C \cap L$ is the subscheme of $\mathbf{A}^2$

defined by the sum of the ideals I_C and I_L, so that, for example, the line $(y = 0)$ and the parabola $(y = x^2)$ will intersect in the subscheme $X_{0,1} = \operatorname{Spec} k[x, y]/(x^2, y)$.

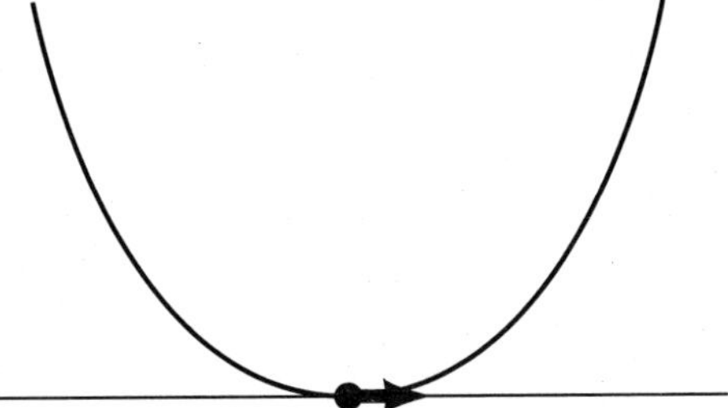

Note that this does indeed determine L: L is the unique line in the plane containing $X_{0,1}$.

a Flatness I: Limits of Pairs of Points

Another important way in which subschemes such as $X_{\alpha, \beta}$ arise is as *limits* of reduced schemes. For example, consider a pair of distinct closed points $(0, 0)$ and (a, b) in the plane. Their union is the closed subscheme

$$X = \{(0,0), (a, b)\} = \operatorname{Spec} S \subset \mathbf{A}^2$$

with

$$\begin{aligned}
S &= k[x, y]/((x, y) \cap (x - a, y - b)) \\
  &= k[x, y]/(x^2 - ax, xy - bx, xy - ay, y^2 - by)
\end{aligned}$$

By the Chinese remainder theorem, $S \cong k \times k$; so in particular, S is a k-algebra of (vector space) dimension 2 over k.

Now suppose the point (a, b) moves toward the point $(0, 0)$ along a curve $(a(t), b(t))$, with $(a(0), b(0)) = (0, 0)$, where a and b are polynomials in t; we write

$$a(t) = a_1 t + a_2 t^2 + \cdots$$

and

$$b(t) = b_1 t + b_2 t^2 + \cdots$$

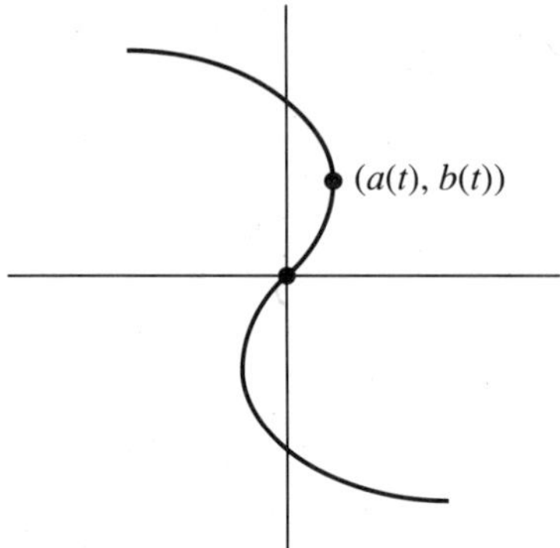

What should be the limit of $X_t = \{(0,0), (a(t), b(t))\}$ as $t \to 0$? Using schemes, we can afford the luxury of the idea that it will continue to be two points, in a suitable sense: it will be an affine scheme X whose coordinate ring is again a two-dimensional vector space over k. We may define X by taking its ideal to be the limit as $t \to 0$ of the ideal $I_t = (x, y) \cap (x - a(t), y - b(t))$. Of course, this only shifts the burden to describing what is the limit of a family of ideals! But this is easy: in the current case, for example, we can take their limit as codimension 2 subspaces of $k[x, y]$, viewed as a vector space over k. That this limit is again an ideal follows from the continuity of multiplication. A more delicate description is necessary in the general case, where the ideals are of infinite codimension; we will discuss this below when we come to limits of families of one-dimensional schemes and again in Section III-B in the projective case.

To see what this means in practice, observe first that the generators $x^2 - a(t)x$, $xy - b(t)x$, $xy - a(t)y$, and $y^2 - b(t)y$ of the ideal I_t clearly have as limit when $t \to 0$ the polynomials x^2, xy, xy, and y^2, so these will be in I. In addition, observe that I_t contains the linear form

$$a(t)y - b(t)x = (xy - b(t)x) - (xy - a(t)y)$$

and hence for $t \neq 0$ the polynomial

$$\frac{a(t)y - b(t)x}{t} = a_1 y - b_1 x + t(\ldots)$$

The ideal I thus contains the limit $a_1 y - b_1 x$ of this polynomial as well; so we have $I \supset (x^2, xy, y^2, a_1 y - b_1 x)$. But the right-hand side of this expression already has codimension 2 as a vector subspace in the polynomial ring $k[x, y]$. Thus

$$I = (x^2, xy, y^2, a_1 y - b_1 x)$$

and correspondingly

$$\lim_{t \to 0} (X_t) = X_{\alpha, \beta}$$

with $\alpha = b_1$, $\beta = -a_1$.

From this we see that X, as a subscheme of $\mathbf{A}_k^2$, "remembers" the direction of approach of $(a(t), b(t))$; we think of it as consisting of the origin together with a tangent vector pointing in this direction, along the line with equation $a_1 y - b_1 x = 0$. This line is the limit of the lines L_t joining $(0,0)$ to $(a(t), b(t))$; that is, it is the tangent line to the curve parametrized by $(a(t), b(t))$ at the origin:

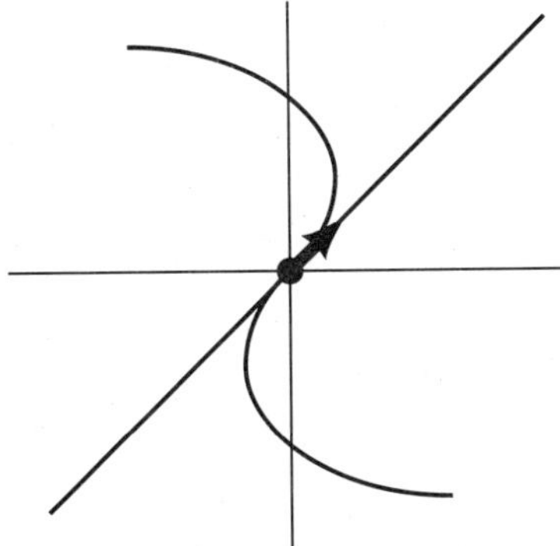

We will see how to generalize this notion of limit in II-C-iii-b (where we will also explain the title of this section).

ii Multiple Points

The subschemes $X_{\alpha, \beta}$ of the preceding examples are called "double points" in the plane, the *double* referring to the vector space dimension of their coordinate rings $R = k[x, y]/(x^2, xy, y^2, \alpha x + \beta y) \cong k[t]/(t^2)$ as k-modules. In general, if $X = \operatorname{Spec} R$ is an affine scheme and R is a finite-dimensional k-vector space, we define the degree of X to be the dimension of R. We next consider examples having degree greater than or equal to 3.

A number of things are different here. To begin with, all double points over an algebraically closed field k—that is, schemes of the form $\operatorname{Spec} R$, where R is a local k-algebra of length 2—are isomorphic, since the only commutative local algebra of vector space dimension 2 over an algebraically closed field k is $k[x]/(x^2)$. (Proof: Let $\mathcal{O}$ be the local ring. Since there is no finite-dimensional extension field of k, the residue class field of $\mathcal{O}$ is k; and since $\mathcal{O}$ is two-dimensional, this implies that its maximal ideal is one-dimensional. Of course, its maximal ideal must have square 0 (for example, by Nakayama's lemma), so the obvious map from $k[x]$ onto $\mathcal{O}$ has x^2 in the kernel and identifies $\mathscr{O}$ with $k[x]/(x^2)$ as required.) By contrast, this is not true of triple points: the schemes

$$X = \operatorname{Spec} k[x]/(x^3)$$

and

$$Y = \operatorname{Spec} k[x, y]/(x^2, xy, y^2)$$

are readily seen to be nonisomorphic. However, any triple point is isomorphic to either X or Y, a fact whose proof we leave as the following exercise.

EXERCISE II-9*

Let $Z = \operatorname{Spec} k[x_1, \ldots, x_n]/I \subset \mathbf{A}_k^n$ be any subscheme of dimension 0 and degree 3, supported at the origin. Show that Z is isomorphic either to the scheme $X \cong \operatorname{Spec} k[x]/(x^3)$ or to $Y \cong \operatorname{Spec} k[x, y]/(x^2, xy, y^2)$ above, and that these are not isomorphic to one another.

In particular, any $k[x_1, \ldots, x_n]/I$ of vector space dimension 3 over k can be generated by two linear forms in the x_i. In geometric terms, any triple point in $\mathbf{A}_k^n$ is planar—that is, lies in a linear subspace $\mathbf{A}_k^2 \subset \mathbf{A}_k^n$. Inside $\mathbf{A}_k^2$ both types of triple points can be realized as limits of triples of distinct points. The ones isomorphic to X above may be obtained from three points coming together in the plane along a smooth curve, while those isomorphic to Y above arise when two points approach a third from different directions. The following exercises contain examples of these phenomena.

EXERCISE II-10

Show that the subscheme of $\mathbf{A}^2$ given by the ideal $(y - x^2, xy)$ arises as the limit of three points on the conic curve $y = x^2$ and is isomorphic to X above, but is not contained in any line in $\mathbf{A}^2$.

Show also that subschemes of $\mathbf{A}^2$ isomorphic to Y above arise when two points approach a third from different directions.

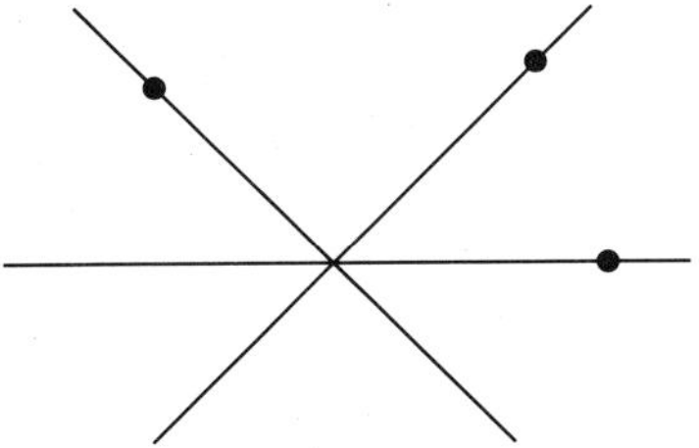

EXERCISE II-11

(For those familiar with the Grassmannian.) The examples above may lead one to expect that the schemes isomorphic to X are limits of those isomorphic to Y. In fact, just the opposite is the case, in the following sense. Let $\mathscr{H}$ be the set of such subschemes supported at the origin; this is naturally parametrized by a closed subscheme of the Grassmannian of codimension 3 subspaces of the six-dimensional vector space $k[x,y]/(x,y)^3$. Show that $\mathscr{H}$ is a surface, with one point corresponding to the unique subscheme $\operatorname{Spec} k[x,y]/(x^2,xy,y^2)$ isomorphic to Y and the rest corresponding to subschemes isomorphic to X. Show that the scheme $\mathscr{H}$ is isomorphic to a two-dimensional quadric cone in $\mathbf{P}^3$, and that the vertex is the one point corresponding to Y.

EXERCISE II-12

Consider the subschemes

$$X_t = \{(0,0,0,\ldots),(t,t^2,t^3,\ldots),(2t,4t^2,8t^3,\ldots)\} \subset \mathbf{A}_k^n$$

defined for $t \neq 0$ on the curve C given by the ideal

$$J = (x_2 - x_1^2, x_3 - x_1^3, \ldots)$$

(C is the image of the map $\mathbf{A}_k^1 \to \mathbf{A}_k^n$ given by

$$t \to (t,t^2,t^3,\ldots,t^n))$$

a. Show that the limit scheme as $t \to 0$ is

$$X_0 = \operatorname{Spec} k[x_1,\ldots,x_n]/(x_2 - x_1^2, x_1 x_2, x_3, x_4, \ldots, x_n)$$

and is isomorphic to the triple point $\operatorname{Spec} k[x]/(x^3)$ above.

b. Show, however, that X_0 is not contained in the tangent line to C at the origin! Rather, the smallest linear subspace of $\mathbf{A}_k^n$ in which X_0 lies is the osculating 2-plane

$$x_3 = x_4 = \cdots = x_n = 0$$

to C (recall that this is by definition the limit of the planes spanned by the tangent line and another point on C near the origin as the point approaches the origin), while the tangent line to C is the smallest linear subspace of $\mathbf{A}_k^n$ containing the subscheme defined by the square of the maximal ideal in the coordinate ring of X_0. Thus, in this sense, X_0 "remembers" both the tangent line and the osculating 2-plane to C.

EXERCISE II-13

Consider the subschemes

$$X_t = \{(0,0), (t,0), (0,t)\}$$

defined for $t \neq 0$, each consisting of three distinct points in $\mathbf{A}_k^2$.

a. Show that the limit scheme as $t \to 0$ is

$$X_0 = \operatorname{Spec} k[x, y]/(x^2, xy, y^2)$$

b. Show that the restriction of a function $f \in k[x, y]$ on $\mathbf{A}_k^2$ to X_0 determines and is determined by the values at the origin of f and its first derivatives in every direction; thus we think of it as a first-order infinitesimal neighborhood of the point $(0,0)$.

c. Show that X_0 is contained in the union of any two distinct lines through $(0,0)$.

d. Show that X_0 is not contained in any smooth curve and thus, in particular, is not the scheme-theoretic intersection of any two smooth curves in $\mathbf{A}_k^2$.

As we said, both types of triple point are contained in planes inside any affine space in which they are embedded. But the 4-tuple point $\operatorname{Spec} k[x, y, z]/(x, y, z)^2$ is not, since its maximal ideal cannot be generated by two elements. Other new phenomena occur for spatial multiple points— those not contained in the plane—and those in still higher-dimensional spaces. For example, not every 8-tuple point in 4-space arises as a limit of sets of eight distinct points! (See, for example, Iarrobino [1985].)

EXERCISE II-14*

Classify subschemes of $\mathbf{A}_k^2$ of dimension 0 and degrees 4 and 5 with support at the origin up to isomorphism. Which are isomorphic as schemes over $\operatorname{Spec} k$?

EXERCISE II-15

A scheme $\operatorname{Spec} R$ supported at a point is called *curvilinear* if the maximal ideal of the (necessarily local) ring R is generated by one element. (The name comes from the fact that these are exactly the schemes that can be contained in a smooth curve.) Show that any two curvilinear subschemes of $\mathbf{A}^2$ having degree 2 and supported at a point can be transformed into one another by a linear transformation of the plane, but that this is not possible for curvilinear schemes of length 3. (Note, however, that any two curvilinear subschemes of $\mathbf{A}^2$ of the same degree *can* be carried into one another by an automorphism of $\mathbf{A}^2$.)

EXERCISE II-16*

(For those with some familiarity with curves.) There are infinitely many isomorphism types of degree 7 subschemes supported at the origin in 3-space and infinitely many types of degree 8 subschemes supported at the origin in the plane.

The following exercise gives an example of the behavior of nonreduced schemes over non-algebraically closed fields.

EXERCISE II-17

Classify all schemes of degree 3 supported at the origin in $\mathbf{A}^2_{\mathbf{R}}$. In particular, show that while any such scheme X whose complexification $X \times_{\operatorname{Spec} \mathbf{R}} \operatorname{Spec} \mathbf{C}$ is isomorphic to $\operatorname{Spec} \mathbf{C}[x]/(x^3)$ is itself isomorphic to $\operatorname{Spec} \mathbf{R}[x]/(x^3)$, there are exactly two nonisomorphic schemes X whose complexification is isomorphic to $\operatorname{Spec} \mathbf{C}[x, y]/(x^2, xy, y^2)$.

iii Embedded Points

We consider now some examples of nonreduced schemes of higher dimension; for simplicity we will restrict ourselves to the case where the underlying reduced scheme is a line. Even so, the variety of possible behaviors increases enormously; for example, we can have schemes that look like reduced schemes except at a point or schemes that are everywhere nonreduced. In this subsection, we consider the former type. By way of terminology, we will say that a scheme $X = \operatorname{Spec} k[x_1, \dots, x_n]/I \subset \mathbf{A}^n_k$ has an *embedded component* if for some open subset $U \subset \mathbf{A}^n_k$ meeting X the closure of $X \cap U$ does not equal X; or if, equivalently, the primary decomposition of the ideal I contains embedded primes. See the discussion of primary decomposition that follows. If the embedded prime is maximal—or, equivalently, the open set U may be taken to be the complement of a point—we talk about an *embedded point*; since the schemes X we will discuss below are all one-dimensional, this is all we will see.

The simplest example of a nonreduced scheme that is reduced except at one point is $X = \operatorname{Spec} k[x,y]/(y^2, xy) \subset \mathbf{A}_k^2$. The ideal $I = (y^2, xy) \subset k[x,y]$ is the ideal of functions on the plane vanishing along the line $(y = 0)$ and in addition vanishing to order 2 at the point $(0,0)$; in algebraic terms, this means that $(y^2, xy) = (y) \cap (x,y)^2$. We can thus think of the scheme X as the line $(y = 0)$ with the proviso that a function f on X is defined by its restriction $f(x,0)$ to the line $y = 0$ together with the specification of its normal derivative at the point $(0,0)$—that is, together with the number $\partial f/\partial y(0,0)$.

It is convenient to realize X as the union of the line defined by $y = 0$ with a nonreduced point—for example, the "first-order neighborhood of the origin" defined by the ideal (x^2, xy, y^2).

—————————●—————————

Such decompositions, called *primary decompositions*, exist for any scheme: we briefly review the background from algebra. For more details see, for example, Atiyah and MacDonald (1969) or, for perhaps the gentlest treatment of all, Northcott (1953).

a Primary Decomposition

Given any ideal I in a polynomial ring (or indeed in any Noetherian ring) R, we define the *associated prime ideals of I* to be the prime ideals $\mathfrak{p}$ such that $\mathfrak{p}$ is the annihilator of some element of R/I. These primes make up a finite set.

An ideal $\mathfrak{q} \subset \mathfrak{p}$ is called *primary* to $\mathfrak{p}$ if $\mathfrak{p}$ is the radical of $\mathfrak{q}$ and for any elements, f, g in R with $fg \in \mathfrak{q}$ but $f \notin \mathfrak{p}$ we have $g \in \mathfrak{q}$; equivalently, $\mathfrak{q}$ is $\mathfrak{p}$-primary if $\mathfrak{p}$ is its radical and the localization map $R/\mathfrak{q} \to R_\mathfrak{p}/\mathfrak{q}R_\mathfrak{p}$ is a monomorphism.

Any ideal I may be expressed as the intersection of primary ideals. Since the intersection of ideals primary to a given prime ideal is again primary to that prime, I can even be expressed as an intersection of ideals that are primary to distinct prime ideals. If this is done in such a way that none of the primary ideals can be left out, the expression is called a *primary decomposition of I*. The primary ideals involved are called *primary components* of I.

The associated primes of I are exactly the radicals of the primary components. The primary component of I having a radical equal to a given associated prime is determined uniquely by I if that prime is minimal among the associated primes (such primary components are called *isolated* components), but in general it is far from unique.

Example Taking $I = (y^2, xy)$ as above, the decomposition

$$I = (y) \cap (x, y)^2$$

already given expresses I as an intersection of primary ideals (the first is prime, the second is primary to (x, y)).

Since neither (y) nor $(x, y)^2$ can be omitted from this expression, it is a primary decomposition and the associated schemes of X (as defined below) are precisely the line X_{red} and the reduced point at the origin, which is Y_{red}.

The primary component associated to (x, y) in the decomposition is not unique; it could have been taken to be (x, y^2), or $(x + y, y^2)$, or indeed any of an infinite number of other such ideals; as well as with their intersection (x^2, xy, y^2), or for that matter with any (x^n, xy, y^2), ... Of course, the primary component (y) corresponding to (y) is unique, because X_{red} is not contained in any other associated scheme.

Despite this nonuniqueness, there is a well-defined *length* for the primary component corresponding to a given associated prime $\mathfrak{p}$, which may be computed, without choosing a primary decomposition, as the length of the largest ideal of finite length in the ring $R_{\mathfrak{p}}/IR_{\mathfrak{p}}$. Here the *length* of a module is simply the length of any finite composition series for that module with simple factors, if such exists, or infinity in the contrary case.

EXERCISE II-18* ——————————————————————————————

The length of the primary component of (xy, y^2) at the origin is 1.

It is easy to translate these matters into the geometry of schemes; any affine scheme X is the union of "primary" closed subschemes, called *primary components*, where a *primary affine scheme* is an affine scheme Y such that Y_{red} is irreducible and such that if f, g are functions on Y such that

f does not vanish on Y_{red}

but

fg vanishes on Y

then

g vanishes on Y.

In such a *primary decomposition* of X, the components that are (set-theoretically) maximal are unique; and while the others are not unique, the decomposition does have (at least) two nice uniqueness properties:

1. The set of reduced subschemes associated to primary components in a minimal primary decomposition is unique; this is called the set of associated schemes to X.
2. The "length" of the primary component associated to each of the associated schemes of X, called the *multiplicity of that associated scheme* in X, is unique.

By way of terminology, the primary components of X that are set-theoretically maximal are usually called *isolated* components, while the rest are called *embedded* components (because they are embedded in larger components).

We may use our example $X = \operatorname{Spec} k[x, y]/(y^2, xy)$ to illustrate these notions: we have already observed that X is the union of the line

$$X_{\mathrm{red}} = \operatorname{Spec} k[x, y]/(y)$$

and the multiple point

$$Y := \operatorname{Spec} k[x, y]/(x^2, xy, y^2)$$

and we have seen that this gives a primary decomposition, the multiplicity of the embedded subscheme at the origin being 1.

Note that we can write X in many different ways as the union of a line and a point: for any $\alpha \neq 0$, we have $X = Y \cup Z$, where $Z = X_{\mathrm{red}} = \operatorname{Spec} k[x, y]/(y)$ is the line and Y is the subscheme $X_{1,\alpha}$ introduced in the first section of this chapter—that is,

$$Y = \operatorname{Spec} k[x, y]/(x^2, xy, y^2, x + \alpha y)$$

Choosing two such subschemes Y, Y' gives an example of closed subschemes Y, Y' and Z in $\mathbf{A}^2$ such that

$$Y \cup Z = Y' \cup Z \qquad \text{and} \qquad Y \cap Z = Y' \cap Z$$

but $Y \neq Y'$.

In the example above, X can be described as the unique subscheme of $\mathbf{A}^2$ consisting of the (reduced) x-axis plus an embedded point of multiplicity 1 at the origin. But embedded points can carry geometric information, too.

EXERCISE II-19*

Choose a linear embedding of $\mathbf{A}^2$ in $\mathbf{A}^3$, let P be the image of $\mathbf{A}^2$, and let X' be the image of X. Show that X' determines P among all planes in $\mathbf{A}^3$.

It is also interesting to consider subschemes of $\mathbf{A}_k^2$ and $\mathbf{A}_k^3$ supported on a union of two given lines, with an embedded point of multiplicity 1 at the intersection of the two lines. In the plane, if we take the two lines to be the coordinate axes, such a scheme may be given as

$$X = \operatorname{Spec} k[x, y]/(x^2 y, xy^2)$$

Geometrically, this may be viewed as the union of the two lines defined by $xy = 0$ with the point $\operatorname{Spec} k[x, y]/(x^3, x^2 y, xy^2, y^3)$. In 3-space, if we take the lines to be $(x = z = 0)$ and $(y = z = 0)$, we can get such a scheme either as

$$X = \operatorname{Spec} k[x, y, z]/(z, x^2 y, xy^2)$$

that is, the image of the scheme above under the embedding of $\mathbf{A}_k^2$ into $\mathbf{A}_k^3$ as the plane $(z = 0)$—or as

$$Y = \operatorname{Spec} k[x, y, z]/(z^2, xz, yz, xy)$$

This last is just the union of the two lines with the subscheme of $\mathbf{A}_k^3$ defined by the square of the maximal ideal of the origin in $\mathbf{A}_k^3$.

EXERCISE II-20

Prove that $X \not\cong Y$ and that X (respectively Y) is, up to isomorphism, the unique example contained in a plane (respectively contained in 3-space but not in any plane) of two lines meeting in a point and having an embedded point of multiplicity 1 at that intersection point.

One justification for the idea that the multiplicity of the embedded point at the origin in the scheme

$$X = \operatorname{Spec} k[x, y]/(xy, y^2)$$

above is 1 is that X is the limit as $t \to 0$ of the family of subschemes

$$X_{\text{red}} \cup Y_t$$

where Y_t is the scheme consisting of one reduced point

$$\operatorname{Spec} k[x, y]/(x, y - t) \subset \mathbf{A}_k^2$$

This is plausible since the ideal

$$(x, y - t) \cap (y) = (xy, y^2 - ty)$$

of $X_{\mathrm{red}} \cup Y_t$ naturally seems to approach (xy, y^2) at $t \to 0$. However, the notion of limit that we introduced earlier is not quite strong enough to deal with this example, since the ideal $(x, y - t) \cap (y)$ of $X_{\mathrm{red}} \cup Y_t$ is not of finite codimension. We will now describe the general context for taking limits of schemes; we will not prove every assertion here, but we will return to proofs in a case strong enough to deal with the limit above and with the others that we will encounter in the rest of this book.

b Flatness II: Flat Families of Schemes

The definition of a "family of schemes" is extremely general: a family of schemes is simply a morphism $\pi : X \to B$ of schemes! The individual schemes "in the family" are the fibers of π over points of B. This notion includes all others that one can think of, such as "a scheme with parameters in the equations" (B is the space on which the parameters vary). We will define a *family of closed subschemes of a given scheme A* to be the restriction of the projection map $A \times B \to B$ to a closed subscheme $X \subset A \times B$; then the fiber of X over $b \in B$ is naturally a closed subscheme of $A_b = A \times b$.

However, such a notion of family is so general as to be virtually useless, because the fibers of the family may have *nothing* in common. For example, given a family π as above, and a closed point $b \in B$, one could make a new family by replacing X by the disjoint union of $X - \pi^{-1}b$ and some other scheme Y, sending all of Y to b. Thus we must put some condition on π if we wish to have families of schemes that "vary continuously" in some reasonable sense. What *reasonable* should mean is not obvious, but it seems natural at least to as for it to include that great ancestor of all continuously varying families, the family of projective plane curves of a given degree,[2] and also that it include the families of schemes defined by families of ideals of constant finite codimension in a polynomial ring that we considered in the context of limits of multiple points.

In many geometric theories one gets the right notion of a continuously varying family by demanding some local triviality of the family; that is, locally in some suitable sense, π should look like the projection to one factor of a direct product. If we do this naively for schemes, interpreting *locally* as meaning locally in the Zariski topology, we get a notion that is far too

---

[2] The easiest way to make this into a family in our sense is to consider it as the projection

$$\pi : \Gamma \subset \mathbf{P}^2 \times \mathbf{P}^N \to \mathbf{P}^N$$

where $\mathbf{P}^N$ is the space of all homogeneous forms of a given degree and Γ is the *incidence correspondence*, $\Gamma = \{(x, f) \mid f(x) = 0\}$. This is, of course, a description as classical varieties, not as schemes; but after reading Chapter III the reader will have no trouble converting it.

restrictive to be much use. A more sophisticated approach would be to demand this local triviality analytically; that is, to demand that if $x \in X$ and $b = \pi(x)$, then the completion of the local ring $\mathcal{O}_{X,x}$ should look like a power series ring over the completion of the local ring $\mathcal{O}_{B,b}$. This notion is quite useful (it is called *smoothness*), but it excludes, for example, the family of plane curves of a given degree, since a smooth family cannot have singular fibers. Smoothness also excludes the families treated in the previous section, in which a disjoint union of distinct points approaches a multiple point; at the multiple point, the criterion is not met. Thus we must look for something more general.

The best current candidate for such a general notion is that of *flatness*. The definition is easy, though perhaps initially opaque.

First, a module M over a ring R is *flat* if for every monomorphism of R-modules $A \rightarrowtail B$, the induced map $M \otimes_R A \to M \otimes_R B$ is again a monomorphism. In particular, any free module is flat; and thus if R is a field, every module is flat. It is not hard to show that if R is a Dedekind domain, then M is flat iff M is torsion-free.

A family $\pi : X \to B$ of schemes is *flat* if for every point $x \in X$ the local ring $\mathcal{O}_{X,x}$ regarded as an $\mathcal{O}_{B,\pi(x)}$-module via the map π^*, is flat. This notion is general enough to include the families of plane curves of given degree but restrictive enough so that the varieties in a flat family have a lot in common. It is really quite satisfactory, except for the fact that it does not seem to have any very "geometric" description. See Matsumura (1986), Grothendieck (1957–1962), or Hartshorne (1977) for good technical discussions.

A valuable feature of the definition is that it is even satisfactory in case the "base" space B is a scheme such as $\operatorname{Spec} k[\varepsilon]/(\varepsilon^n)$. In this case the deformation is called an "infinitesimal deformation." Such things played an important role in the algebraization of the theory of curves on surfaces—see, for example, the book of Mumford (1966) and the discussion in Section IV-B-vi.

Another thing that makes flatness a good notion is the *generic flatness theorem* due to Grothendieck (1957–1962). This says that if one has any reasonable family of schemes $X \to B$, then there is an open dense subset U of B such that the restricted family $\pi^{-1}U \to U$ is flat (here *reasonable* includes, for example, any family of subschemes of a fixed affine or projective space). In some sense this vindicates our choice of flatness as the analogue of the notion of bundle in topology: it is analogous to the observation that if $f : M \to N$ is a differentiable map of compact $\mathscr{C}^\infty$ manifolds, then there is a dense collection of open subsets U of the target space N such that the restriction of f to each $f^{-1}(U)$ is a fiber bundle. In any event, the generic flatness theorem certainly assures us that flat families are ubiquitous in algebraic geometry.

The most elementary examples of families of varieties, such as those considered in the previous section and those to be considered below, are families where the total space X is affine and the base B is affine and very simple. By Theorem I-36, whenever $B = \operatorname{Spec} R$ is affine, the family π is given by a ring homomorphism $R \to \mathcal{O}_X(X)$.

In our cases, we will have $B = \operatorname{Spec} R$ with $R = k[t]$, $R = k[t]_{(t)}$, or $R = k[[t]]$, one of three versions of a line. Here, fortunately, flatness is easy to understand: $\mathcal{O}_{X,x}$ regarded as an $\mathcal{O}_{B,\pi(x)}$-module is flat for all x iff $\mathcal{O}_X(X)$ is torsion-free as an R-module [see Bourbaki (1989, I.2.4, Proposition 3.ii); because all these rings R are principal ideal domains, this also follows easily from Matsumura (1986, Theorem 7.6 and its converse on p. 50)]. This torsion-freeness condition is automatically satisfied, for example, if X is irreducible and reduced and the map $X \to B$ is onto, since then the map $R \to \mathcal{O}_X(X)$ is an inclusion into a domain. On the other hand, families like the "bad" family considered above,

$$X' = X - \pi^{-1}0 \cup Y \to B$$

where $0 = [(t)]$ is the closed point of B at the origin and Y maps to 0, are visibly not flat because $\mathcal{O}_{X'}(X') = \mathcal{O}_{X-\pi^{-1}0} \times \mathcal{O}_Y(Y)$, and the elements of $\mathcal{O}_Y(Y)$ are all annihilated by t.

EXERCISE II-21

Show that the map in part a of Exercise I-39 makes X into a flat family of schemes over Y. By contrast, consider the map given in part b. Let L be an arbritrary line through the origin in the plane Y, and let M be the (scheme-theoretic!) preimage of L in X. Set-theoretically, M is the union of two lines, like the example in part a. Show that as a scheme, it is actually nonreduced, with an embedded point where the lines meet. Conclude that the map from M to L is not flat. If you know some commutative algebra, you may deduce that the original map $X \to Y$ of part b is not flat. Note that as explained in part b, the fiber over the origin consists of a triple point, whereas all the other fibers consist of pairs of points; this is also enough to check nonflatness.

EXERCISE II-22*

(For those with some knowledge of commutative algebra.) Let R be a domain. Show that every flat module is torsion-free. Assuming in addition that R is Noetherian, show that every torsion-free R-module is flat iff R is Dedekind.

From here on, we will deal only with the base $B = \operatorname{Spec} R$, with $R = k[t]_{(t)}$ and X affine, and simply take the torsion-freeness of $\mathcal{O}_X(X)$ as our definition of flatness. Using this definition, we will again begin to prove our assertions.

The scrupulous reader will wish to reconcile these notions with what we did when we studied limits of finite schemes in Section II-C-i-a. Recall that in that discussion we defined the limits differently: we worked with families of schemes over $B = \operatorname{Spec} k[t]$ defined by ideals $I_t \subset k[t][x_1,\ldots,x_n]$ such that the vector space dimension of the algebra $\mathcal{Q}_X(X) = k[t][x_1,\ldots,x_n]/I_t$ was constant for $t \neq 0$, say

$$\dim k[t][x_1,\ldots,x_n]/(t - a, I_t) \equiv d \qquad \text{for } 0 \neq a \in k$$

and we said that I_0, the image of I_t in $k[t][x_1,\ldots,x_n]/(t)$, represented the correct limit iff

$$\dim k[t][x_1,\ldots,x_n]/(t, I_t) = d$$

as well. In the restricted case above, this implies that the family is flat. To see why, we make use of the fact in all the situations we were considering, $\mathcal{O}_X(X)$ was a finitely generated module over $R = k[t]$. Since R is a principal ideal domain, a finitely generated module is torsion-free (and thus flat) iff it is free. But another criterion for the freeness of a finitely generated R-module is this: M is free iff in the family of all maximal ideals $\mathfrak{m}$ (in our case $\mathfrak{m} = (t - a)$, for $a \in k$) of R, the dimension of $M/\mathfrak{m}M$ as a vector space over k is independent of $\mathfrak{m}$. Taking $M = \mathcal{O}_X(X)$, the vector space in question is just

$$k[t][x_1,\ldots,x_n]/(t - a, I_t)$$

whence the desired implication. We will see in the following chapter a vast generalization of this implication, once we introduce the notion of the Hilbert polynomial of a scheme.

Here is another criterion for flatness that is extremely useful in practice.

EXERCISE II-23* ───

Let $A = R[x_1,\ldots,x_n]$ be a polynomial ring over $R = k[t]_{(t)}$, and let M be an A-module with free presentation

$$F_1 \xrightarrow{\varphi} F_0 \to M \to 0$$

Consider the module $\overline{M} := M/tM$ over the factor ring $\overline{A} := A/tA$, and let

$$\overline{F}_1 \xrightarrow{\overline{\varphi}} \overline{F}_0 \to \overline{M} \to 0$$

be the corresponding presentation. Show that M is flat over R iff every second syzygy of $\overline{M}$ over $\overline{A}$ can be lifted to a second syzygy over A in the sense that every element of the kernel of $\bar{\varphi}$ comes from an element of the kernel of φ. (The same is true for any local base ring R with maximal ideal $\mathfrak{m}$ if M is finitely generated over A [or something a bit weaker]; this is a form of the "local criterion of flatness"—see, for example, Matsumura [1986, p. 174].)

It is worth nothing that even in the context of finite schemes, the notion of a flat family of closed subschemes is a little more general than that of Section II-C-i-a. The following exercise shows that the flat limit of a family of closed subschemes consisting of one point each may actually be empty.

EXERCISE II-24 ____

a. Let $X = \operatorname{Spec} k[x]$ and $B = \operatorname{Spec} k[t]$ and let $\Delta \subset X \times B$ be the "diagonal" subscheme defined by the ideal $(x - t)$. The projection map $\varphi : \Delta \to B$ is an isomorphism—thus, in particular, flat—and thus the limit, as $t \to 0$, of the one-point subschemes of X that are the fibers over points of B other than 0 is the one-point subscheme of X. Check that this is the limit in the sense of Section II-C-i-a as well.

b. Now, let $X' = X - \{0\} = \operatorname{Spec} k[x, x^{-1}]$, and let $\Delta' = \Delta \cap (X' \times B)$. Show that the induced map $\Delta' \to B$ is also flat, and therefore the limit as $t \to 0$ of the one-point closed subscheme of Δ' is the empty subscheme. Check that this limit does not satisfy the condition of Section II-C-i-a for constancy of codimension, and in fact that there is no ideal in the coordinate ring of $X' \times B$ with constant codimension 1 in each fiber. The following picture explains the situation:

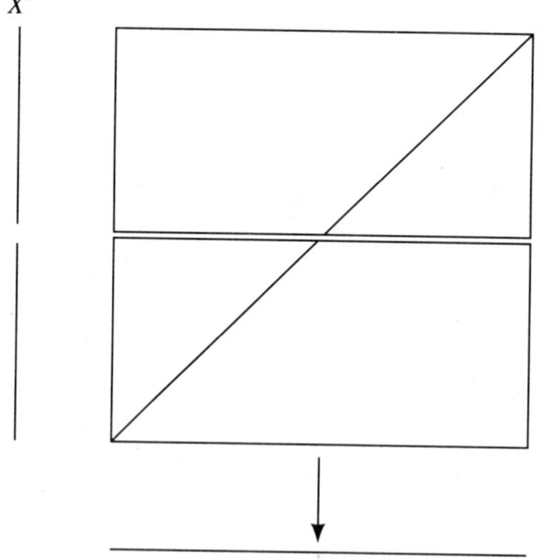

A very attractive feature of flatness that comes from this equivalence of flatness and torsion-freeness over B is that one-parameter families of closed subschemes always have uniquely defined flat limits in the following sense.

PROPOSITION II-25   Let $B = \operatorname{Spec} R$, with $R = k[t]_{(t)}$. Given any family of closed subschemes of an affine space defined over $B - \{[(t)]\}$, there is a unique way of completing it to a flat family of schemes over B; that is, given any ideal I' in $k(t)[x_1, \ldots, x_n]$, there is a unique ideal I in $R[x_1, \ldots, x_n]$ such that

1. $k(t)I = I'$,
2. $R[x_1, \ldots, x_n]/I$ is flat over R.

In fact, I is given by the formula

$$I = I' \cap R[x_1, \ldots, x_n]$$

The *flat limit* as $t \to 0$ of the family defined by I' in the Proposition is, of course, the image of I in $k[x_1, \ldots, x_n] = R[x_1, \ldots, x_n]/(t)$.

Proof   If $I = I' \cap R[x_1, \ldots, x_n]$, then $R[x_1, \ldots, x_n]/I$ is torsion-free, and thus flat, because it is a submodule of $k(t)[x_1, \ldots, x_n]/I'$.

This ideal I is obviously the largest ideal of $R[x_1, \ldots, x_n]$ such that $I \cdot k(t) = I'$. If J is another, then $J \subset I$; and because they become equal over $k(t)$, the quotient I/J must be torsion. But $I/J \subset R[x_1, \ldots, x_n]/J$, so this module is not flat.   □

Any ideal I satisfying 1 above corresponds to a family of schemes over B "completing" the given family. The point here is that there is a unique way to complete the family—which is trivially flat since $B - \{[(t)]\} = \operatorname{Spec}(k[t]_{(t)})_t = \operatorname{Spec} k(t)$ is the spectrum of a field—to a flat family over B.

EXERCISE II-26 ──

Formulate and prove the analogous assertion for the case when $R = k[[t]]$ and in the case $R = k[t]$.

EXERCISE II-27 ──

Show that in all three cases we can replace the affine space with an arbitrary fixed scheme.

EXERCISE II-28 ──

Two-parameter families may not admit any flat limits at all! For example, consider the family $\Gamma \to \mathbf{A}^2 - \{(0,0)\}$ given by projection on the second factor from

$$\Gamma = \{(x, (s, t)) \in \mathbf{A}^1 \times (\mathbf{A}^2 - \{(0,0)\}) \mid sx - t = 0\}$$

Show that there is no flat limit as (s, t) tends to $(0,0)$ by showing that if we restrict to lines through the origin in $\mathbf{A}^2$, every possible limit is obtained.

Applying this procedure in the case $I' = (xy, y^2 - ty)$, the ideal of the union of the line $y = 0$ and the point $x = y - t = 0$ tending to $(0,0)$, whose consideration we began above, we wish to show that the flat limit is given by (xy, y^2). For this it will suffice, by the proposition, to show that $R[x_1, \ldots, x_n]/I'$ is already torsion-free—that is, flat as a $k[t]_{(t)}$-module—or, equivalently, that t is a nonzero divisor modulo

$$(xy, y^2 - ty) = (y) \cap (x, y - t)$$

Quite generally, the set of zero divisors modulo a given ideal is the union of the associated primes of the ideal; since t is not in either of the two associated primes, it is a nonzero divisor, and we are done.

A surprising and important example is the limit of the scheme consisting of two disjoint lines in $\mathbf{A}^3$ as the lines move to meet in a single point; as the following exercise shows, their limit actually has an embedded point at the point of intersection:

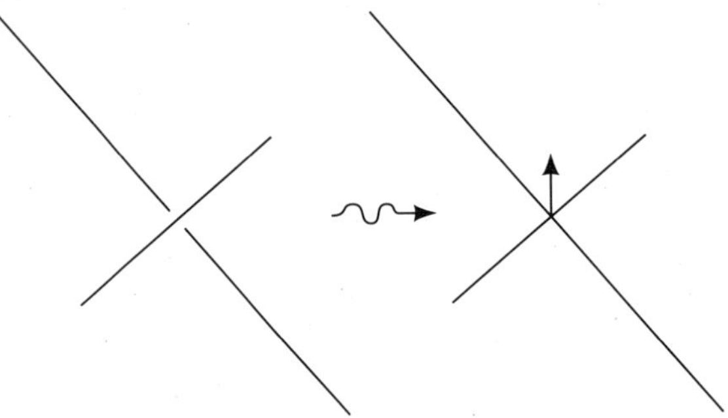

EXERCISE II-29*

Let L_t be the line in $\mathbf{A}^3_k$ defined by the ideal $(y, z - t)$ and M be the line defined by (x, z); for $t \neq 0$ let X_t be their union. Show that the limit of X_t as $t \to 0$ is the scheme

$$Y = \operatorname{Spec} k[x, y, z]/(z^2, xz, yz, xy)$$

Show that there does not exist a family of lines L_t such that the limit of $M \cup L_t$ as $t \to 0$ is the scheme

$$X = \operatorname{Spec} k[x, y, z]/(z, x^2 y, xy^2)$$

Note that in this example (as in others throughout this book) the limit of a union of schemes properly contains the union of their limits.

iv Multiple Lines

We now consider a nonreduced affine scheme X supported on a line and not having embedded components. We will assume that the multiplicity of the line (in the sense of the primary decomposition) is 2, and we will analyze the possibilities.

It is very easy to write down a first example: the scheme

$$X = \operatorname{Spec} k[x, y]/(y^2) \subset \mathbf{A}_k^2$$

obviously has the desired properties. It is pretty clear that there are no more examples that support the line $y = 0$ in $\mathbf{A}_k^2$, but we can construct many in $\mathbf{A}_k^3$. A subscheme X of the sort we want will meet a general plane in $\mathbf{A}_k^3$ passing through a point of the reduced line in a double point contained in that plane. We already know that any double point may be thought of as a point plus a tangent vector at that point, and this suggests that we obtain X by choosing a normal direction at each point of the line. For example, take $L := X_{\mathrm{red}}$ to be the line $x = y = 0$, with coordinate z. Now, choose a pair of polynomials p and q in z without common zeros, and at each point of L take the normal direction to be the one with slope $p(z)/q(z)$ in the normal plane $z = 0$. It is easy to see that the union over all z of the double points in the given directions will be contained in the scheme $X_{p,q}$ defined by taking

$$I_{p,q} = (x^2, xy, y^2, p(z) \cdot x - q(z) \cdot y)$$

and

$$X_{p,q} = \operatorname{Spec} k[x, y, z]/I_{p,q}$$

The simplest nonplanar example would be one where the chosen normal directions twist just once around L—for example, the one given by the ideal

$$I_\Gamma = (x^2, xy, y^2, zy - x)$$

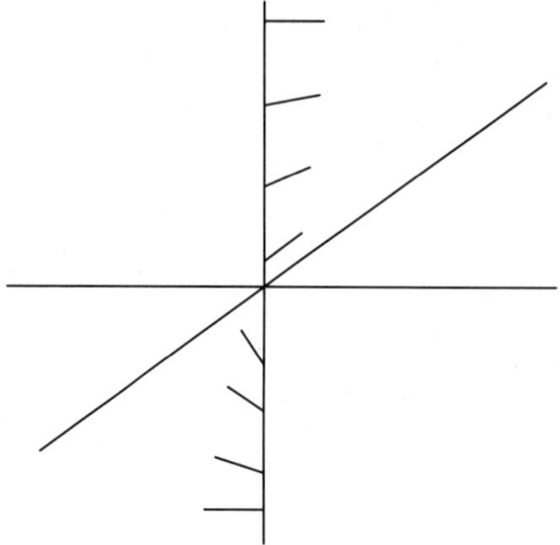

EXERCISE II-30 ____

If p, q are relatively prime polynomials, then the ideal $(x, y)/I_{p,q}$ in the ring $k[x, y, z]/I_{p,q}$ is torsion-free of rank 1 as a $k[z]$-module; and thus $X_{p,q}$ is primary, with $(X_{p,q})_{\text{red}}$ the line $\operatorname{Spec} k[x, y, z]/(x, y)$, and $X_{p,q}$ has multiplicity 2.

At first sight it looks as though these examples will possess many interesting invariants and thus, in particular, be distinct, but this is not so: we can "untwist" any of the schemes $X_{p,q}$ by an automorphism of $\mathbf{A}^3$ to give an isomorphism of it with the planar double line $\operatorname{Spec} k[y, z]/(y^2)$. To do this, note that since p and q have no common zeros, we may write $1 = a \cdot q + b \cdot p$ for some polynomials $a, b \in k[z]$; thus the matrix

$$\begin{pmatrix} a & b \\ p & -q \end{pmatrix}$$

has unit determinant, so the map $\mathbf{A}^3 \to \mathbf{A}^3$ given by

$$(x, y, z) \mapsto (x' := p(z)x - q(z)y, y' := a(z)x + b(z)y, z)$$

is invertible. Again because the matrix is invertible, we have

$$(x, y) = (x', y') \qquad \text{and } (x^2, xy, y^2) = (x'^2, x'y', y'^2)$$

so the ideal of $X_{p,q}$ is $(x, x^2, xy, y^2) = (x, y^2)$, as required.

More generally, it turns out that there is up to isomorphism only one affine double line in the following sense.

EXERCISE II-31* ____

Prove that if $X = \operatorname{Spec} A$ with A a Noetherian k-algebra such that

i. $X_{\mathrm{red}} \cong \mathbf{A}_k^1$,
ii. X has no embedded components,
iii. X has multiplicity 2,

then X is isomorphic to $\operatorname{Spec} k[x, y]/(y^2)$.

We will see in the next chapter that this situation contrasts with the one in projective space: there are many nonisomorphic projective double lines.

D Arithmetic Schemes

Our last collection of examples will be spectra of rings that are finitely generated and reduced but that do not contain any field at all. In general, the spectra of rings finitely generated over $\mathbf{Z}$ are called *arithmetic schemes*; they arise primarily in the context of number theory, although by no means all schemes of number-theoretic interest are of this type. In these examples we will see some hint of the amazing unification that schemes allow between the arithmetic and the geometric points of view.

i Spec Z

We start with the most obvious example, the scheme $\operatorname{Spec} \mathbf{Z}$ itself. The prime ideals of $\mathbf{Z}$ are, of course, the ideals (p), for $p \in \mathbf{Z}$ a prime number, and the ideal (0); the former correspond to closed points of $\operatorname{Spec} \mathbf{Z}$, with residue field $\mathbf{F}_p$, while the latter is a "generic" point, whose closure is all of $\operatorname{Spec} \mathbf{Z}$ and whose residue field is $\mathbf{Q}$. The picture thus looks like this:

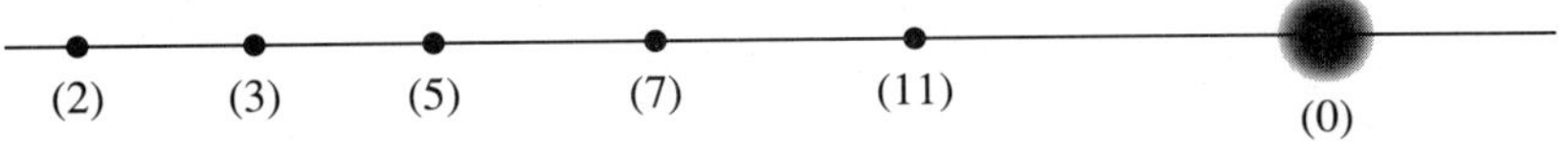

This bears a formal resemblance to an affine line $\mathbf{A}_k^1$ over a field; indeed, this similarity is just the beginning of a long sequence of analogies, and it is well to bear it in mind while looking at the following examples. However, the analogy also has its limits: while $\operatorname{Spec} \mathbf{Z}$ behaves much like $\mathbf{A}_k^1$, for example, it is not an open dense subscheme of any scheme analogous to $\mathbf{P}_k^1$.

ii Spec of the Ring of Integers in a Number Field

Secondly, consider a scheme of the form Spec A, where $A \subset K$ is the ring of integers in a number field K; we will analyze the example $K = \mathbf{Q}[\sqrt{3}]$ and $A = \mathbf{Z}[\sqrt{3}]$. As in the case of Spec $\mathbf{Z}$, there are just two types of points: closed points corresponding to nonzero prime ideals in A, having finite residue fields, and a generic point corresponding to (0) with residue field K. What makes this example interesting is the map Spec $A \to$ Spec $\mathbf{Z}$ induced by the inclusion of $\mathbf{Z}$ in A. Consider, for example, the fiber over a point $[(p)] \in$ Spec $\mathbf{Z}$. This is just the set of primes in A containing the ideal $pA \subset A$, and it may behave in any one of three ways (a good basic reference for the unexplained material here is Serre [1979]):

i. If p divides the discriminant 12 of K over $\mathbf{Q}$—that is, for $p = 2$ or 3—the ideal (p) is the square of an ideal in A: we have

$$2 \cdot A = (1 + \sqrt{3})^2$$

and, of course,

$$3 \cdot A = (\sqrt{3})^2$$

Note that the residue field at the points $(1 + \sqrt{3})$ and $(\sqrt{3}) \in$ Spec A are the fields $\mathbf{F}_2$ and $\mathbf{F}_3$, respectively.

On the other hand, if p does not divide the discriminant of K over A, there are two further possibilities.

ii. If 3 is a square mod p, the prime (p) will factor into a product of distinct primes: for example, we have

$$(11) \cdot A = (4 + 3\sqrt{3}) \cdot (4 - 3\sqrt{3})$$

and

$$(13) \cdot A = (4 + \sqrt{3}) \cdot (4 - \sqrt{3})$$

The residue fields at these points will again be the prime fields, in this case $\mathbf{F}_{11}$ and $\mathbf{F}_{13}$, respectively. Finally,

iii. If 3 is not a square mod p—as is the case when $p = 5$ or 7, for example—then the ideal $(p) \cdot A$ is still prime and corresponds to a single point in Spec A. In these cases, the residue field is the quadratic extension of $\mathbf{F}_p$—for instance, $\mathbf{F}_{25}$ and $\mathbf{F}_{49}$ in the two examples.

In general, as in this example, if K is a quadratic number field, and A is the ring of algebraic integers in K, then the inclusion $\mathbf{Z} \subset A$ induces a map of schemes $\psi : \operatorname{Spec} A \to \operatorname{Spec} \mathbf{Z}$ whose fiber over each closed point $(p) \in \operatorname{Spec} \mathbf{Z}$ is one of the following:

i. A single, nonreduced point, with coordinate ring isomorphic to $A/\mathfrak{p}^2$, whose underlying reduced point $\mathfrak{p}$ has residue field $\mathbf{F}_p$, if p "ramifies" in A—that is, if $p \cdot A$ is the square of a prime ideal $\mathfrak{p}$ of A.
ii. The disjoint union of two reduced points, $\mathfrak{p}$ and $\mathfrak{p}'$, with residue fields $A/\mathfrak{p} = A/\mathfrak{p}' = \mathbf{F}_p$, if $p \cdot A$ is a product of two distinct prime ideals of A.
iii. A single reduced point $\mathfrak{p}$, with residue field $A/\mathfrak{p}$ of degree 2 over $\mathbf{F}_p$, if p remains prime in A.

Note that in every case the coordinate ring of the fiber has dimension 2 as an $\mathbf{F}_p$-algebra; or, as it is usually written in number theory books, $2 = \sum e_i f_i$—that is because A is a free $\mathbf{Z}$-module of rank 2.

Of interest here is the analogy between the map $\operatorname{Spec} A \to \operatorname{Spec} \mathbf{Z}$ and a branched cover of Riemann surfaces (or, more generally, of one-dimensional schemes over an algebraically closed field such as $\mathbf{C}$). Essentially, we may think of $\operatorname{Spec} A$ as a two-sheeted cover of $\operatorname{Spec} \mathbf{Z}$, with branching over the "ramified" primes, just as, for example, $\operatorname{Spec} \mathbf{C}[z]$ is a double cover of $\operatorname{Spec} \mathbf{C}[z^2]$ branched over the origin—that is, the ideal (z^2). The one apparent difference is that over some points $(p) \in \operatorname{Spec} \mathbf{Z}$ other than ramification points we may have, instead of two distinct points with multiplicity 1, one point with multiplicity 1 but with a residue field that is a quadratic extension of the residue field $\mathbf{F}_p$ at (p).

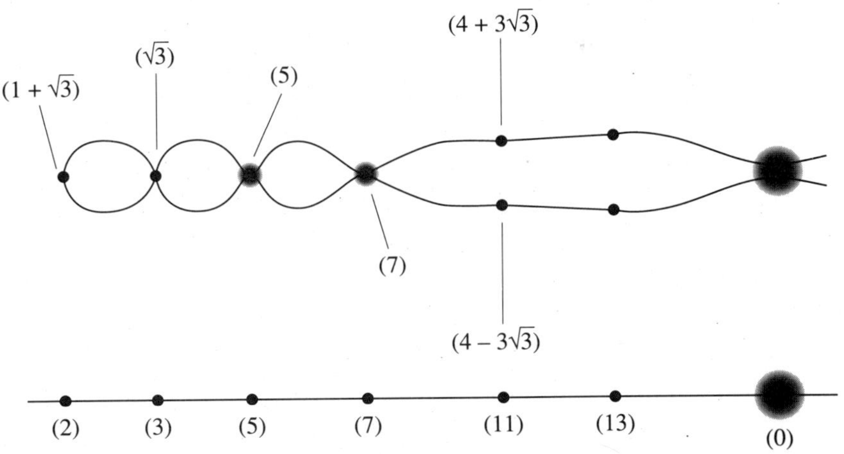

A more inclusive analogy would be with a finite map between one-dimensional schemes over a non–algebraically closed field. Consider, for example, the map

$$\operatorname{Spec} \mathbf{R}[x][y]/(y^2 - x) \to \mathbf{A}^1_{\mathbf{R}} = \operatorname{Spec} \mathbf{R}[x]$$

Looking just at points of $\mathbf{A}^1_{\mathbf{R}} = \operatorname{Spec} \mathbf{R}[x]$ with residue field $\mathbf{R}$—that is, points of the form $(x - \lambda)$ with λ real—we have ramification over the point (x), and for $\lambda \neq 0$ the inverse image of $(x - \lambda)$ is either two distinct points with residue field $\mathbf{R}$ (if $\lambda > 0$) or one point with residue field $\mathbf{C}$ (if $\lambda < 0$).

We may continue this analogy a little further by looking at schemes of the form $\operatorname{Spec} B$, where $B \subset A \subset K$ is an order in a number field, not necessarily equal to the ring of integers. For example, consider the ring $B = \mathbf{Z}[11\sqrt{3}]$ and the associated scheme $\operatorname{Spec} B$. The map $\operatorname{Spec} A \to \operatorname{Spec} \mathbf{Z}$ described above factors through $\operatorname{Spec} B$, and indeed the map $\operatorname{Spec} A \to \operatorname{Spec} B$ is an isomorphism except that the two points $(4 + 3\sqrt{3})$ and $(4 - 3\sqrt{3}) \in \operatorname{Spec} A$ map to the same point $(11) \in \operatorname{Spec} B$. We may thus picture $\operatorname{Spec} B$ as a sort of "nodal curve"—that is, the double cover $\operatorname{Spec} A$ of $\operatorname{Spec} \mathbf{Z}$ with two points identified.

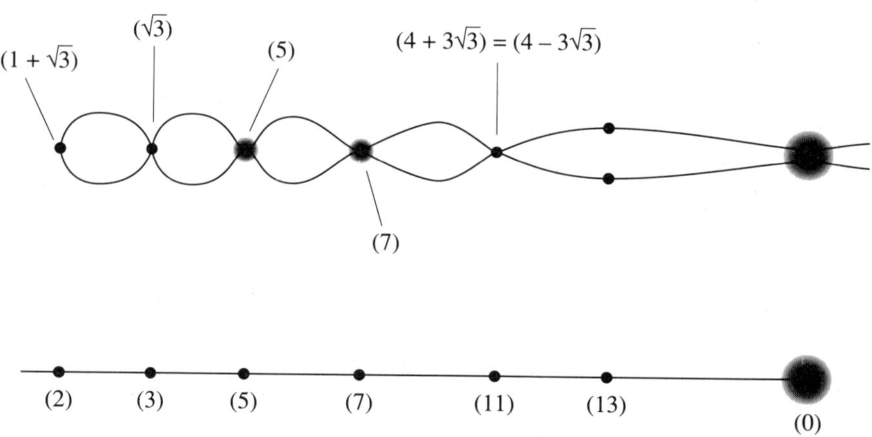

Alternatively, consider the case $B = \mathbf{Z}[2\sqrt{3}]$; here the map $\operatorname{Spec} A \to \operatorname{Spec} B$ is one to one but not an isomorphism at the point $[(1 + \sqrt{3})]$ which goes to $[(2, 2\sqrt{3})]$.

EXERCISE II-32

Show that the point $p = [(2, 2\sqrt{3})]$ is a cusp of the scheme $\operatorname{Spec} \mathbf{Z}[2\sqrt{3}]$ in the sense that it is a singular point and the desingularization $\operatorname{Spec} A \to \operatorname{Spec} B$ has fiber over p consisting of a double point.

iii Affine Spaces Over Spec Z

Our next example is of a two-dimensional scheme, $\operatorname{Spec} \mathbf{Z}[x]$; this is also denoted $\mathbf{A}^1_{\mathbf{Z}}$. The prime ideals in $\mathbf{Z}[x]$ are

   i the ideal (0);

  ii. ideals of the form (p), for $p \in \mathbf{Z}$ prime;

 iii. principal ideals of the form (f), where $f \in \mathbf{Z}[x]$ is a polynomial, irreducible over $\mathbf{Q}$, whose coefficients have greatest common divisor 1;

 iv. maximal ideals of the form (p, f), where $p \in \mathbf{Z}$ is a prime and $f \in \mathbf{Z}[x]$ a monic polynomial whose reduction mod p is irreducible.

EXERCISE II-33*

Prove this.

Of these, only the last are closed points; the first, of course, has closure all of $\mathbf{A}^1_{\mathbf{Z}}$, while the second and third types have closures we will describe below.

Probably the best way to picture $\mathbf{A}^1_{\mathbf{Z}}$ is via the map $\mathbf{A}^1_{\mathbf{Z}} \to \operatorname{Spec} \mathbf{Z}$ (again a flat map!). Under this map, points of type ii and iv above go to the corresponding points $(p) \in \operatorname{Spec} \mathbf{Z}$, while the points of types i and iii go to the generic point $(0) \in \operatorname{Spec} \mathbf{Z}$. Indeed, the fiber of this map over the point (p) is isomorphic to $\mathbf{A}^1_{\mathbf{F}_p} = \operatorname{Spec} \mathbf{F}_p[x]$, with the point $(p, f) \in \mathbf{A}^1_{\mathbf{Z}}$ corresponding to the point in $\mathbf{A}^1_{\mathbf{F}_p}$ given by the set of roots of the polynomial f in the algebraic closure $\overline{\mathbf{F}}_p$ (recall that points of $\mathbf{A}^1_{\mathbf{F}_p}$ correspond to orbits of the action of the Galois group $\operatorname{Gal}(\overline{\mathbf{F}}_p/\mathbf{F}_p)$ on $\mathbf{F}_p$). Similarly, the fiber over the generic point $(0) \in \operatorname{Spec} \mathbf{Z}$ is the scheme $\mathbf{A}^1_{\mathbf{Q}} = \operatorname{Spec} \mathbf{Q}[x]$, with $(f) \in \mathbf{A}^1_{\mathbf{Z}}$ meeting $\mathbf{A}^1_{\mathbf{Q}}$ in the point corresponding to the set of roots of f in $\overline{\mathbf{Q}}$. The picture thus is as follows:

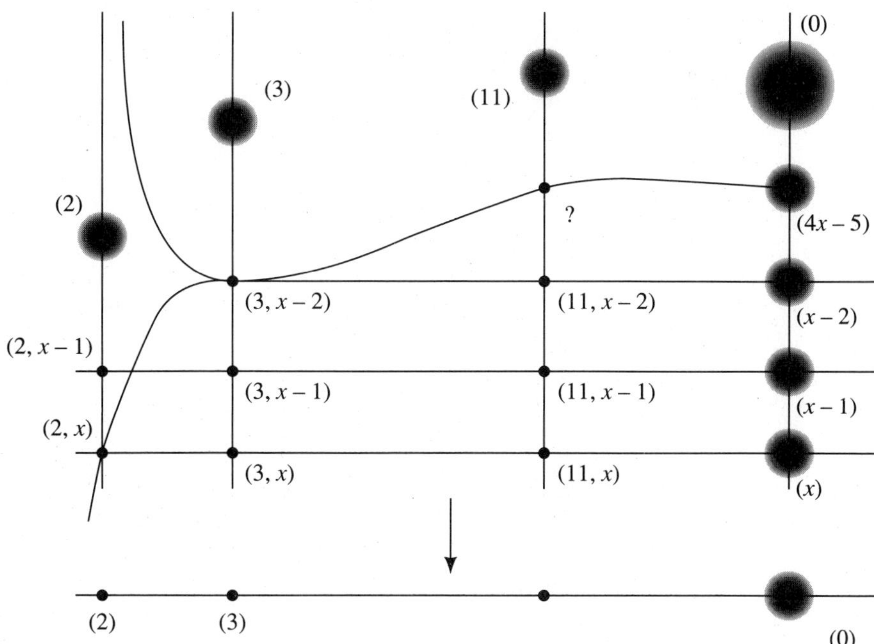

Note that the closure of the point $(p) \in \mathbf{A}^1_{\mathbf{Z}}$ is the fiber $\mathbf{A}^1_{\mathbf{F}_p}$ over the point $(p) \in \operatorname{Spec} \mathbf{Z}$. The closures of the other nonclosed points—those of type iii above—are more interesting. These will consist of the point (f) itself, in the fiber $\mathbf{A}^1_{\mathbf{Q}}$ over (0), together with all the points $(p, g) \in \mathbf{A}^1_{\mathbf{Z}}$, where g is a factor of f over $\overline{\mathbf{F}}_p$—that is, in each fiber $\mathbf{A}^1_{\mathbf{F}_p}$ of $\mathbf{A}^1_{\mathbf{Z}}$, the union of the points of $\mathbf{A}^1_{\mathbf{F}_p}$ corresponding to roots of $f \bmod p$.

EXERCISE II-34

What is the point marked with a ? in the picture above? Why are the closures of the points $(4x - 5)$ and $(x - 2)$ indicated by curves meeting tangentially at the point $(3, x - 2)$, while they are both transverse to the closure of (3)? (See the discussion leading up to Exercise II-40 for one answer.)

For another example, consider the ideal generated by a simple linear polynomial, such as $(5x - 49)$. To continue the analogy between $\operatorname{Spec} \mathbf{Z}$ and the affine line over a field, we can think of the closure of this point as the graph of the function $49/5$ on $\operatorname{Spec} \mathbf{Z}$; this is a function with a simple pole at the point (5) and a double zero at (7). (Note that this curve is tangent to the closed subscheme $(x = 0)$ in $\mathbf{A}^1_{\mathbf{Z}}$, as evidenced by the fact that the intersection of $(x = 0)$ with the subscheme $(5x = 49)$ is not just the point $(7, x)$ but a non-reduced point supported at this point.)

The closure of the point $(x^2 - 3)$ is pictured below in a slightly different style:

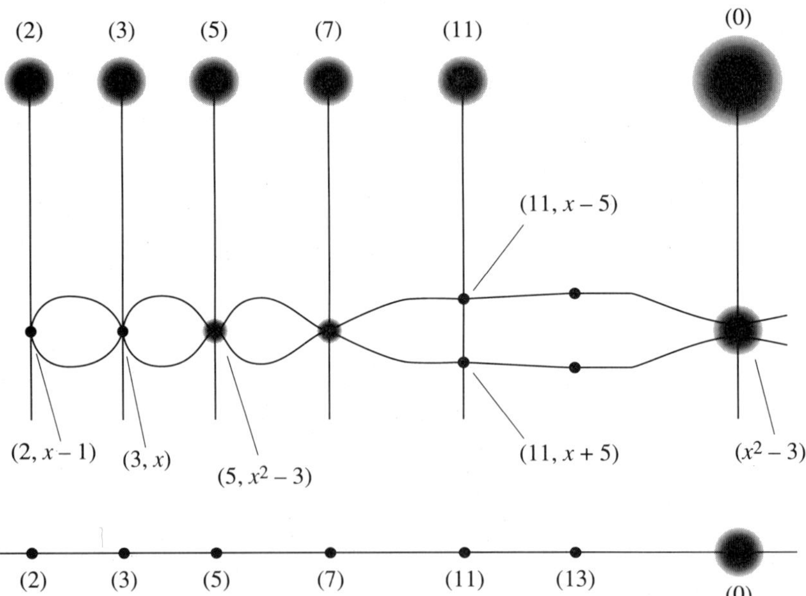

Note that this closure is just the scheme $\operatorname{Spec} \mathbf{Z}[x]/(x^2 - 3) = \operatorname{Spec} \mathbf{Z}[\sqrt{3}]$ described above, realized here as a subscheme of $\mathbf{A}^1_{\mathbf{Z}}$.

EXERCISE II-35 ───

Identify the three unlabeled points in the above diagram.

More generally, the scheme $\mathbf{A}^n_{\mathbf{Z}} = \operatorname{Spec} \mathbf{Z}[x_1, \ldots, x_n]$ can best be viewed via the natural map $\mathbf{A}^n_{\mathbf{Z}} \to \operatorname{Spec} \mathbf{Z}$, whose fibers are the schemes $\mathbf{A}^n_{\mathbf{F}_p}$ and $\mathbf{A}^n_{\mathbf{Q}}$.

iv A Conic Over Spec Z

Our next example gives a hint of the depth of the unification of geometry and arithmetic achieved in scheme theory. We consider the scheme

$$\operatorname{Spec} \mathbf{Z}[x, y]/(x^2 - y^2 - 5)$$

and its morphism to $\operatorname{Spec} \mathbf{Z}$.

To begin with, the fiber of this scheme over the generic point $[(0)] \in \operatorname{Spec} \mathbf{Z}$ is the scheme $X = \operatorname{Spec} \mathbf{Q}[x, y]/(x^2 - y^2 - 5)$, which we have already described: its points are the orbits, under the action of the Galois group $G = \operatorname{Gal}(\overline{\mathbf{Q}}/\mathbf{Q})$, of the set of pairs (x, y) of elements of $\overline{\mathbf{Q}}$ satisfying $x^2 - y^2 = 5$. The fiber over (p) is similarly the subscheme of the affine plane $\mathbf{A}^2_{\mathbf{F}_p}$ over $\mathbf{F}_p$ defined by the equation $x^2 - y^2 = 5$—that is, whose points are the orbits, under the action of the Galois group $G = \operatorname{Gal}(\overline{\mathbf{F}}_p/\mathbf{F}_p)$, of the set of pairs (x, y) of elements of $\overline{\mathbf{F}}_p$ satisfying $x^2 - y^2 = 5$.

Note that the fibers of this scheme over all primes other than 2 and 5 are nonsingular conics, as is the fiber over the generic point.

EXERCISE II-36 ───

Are there plane conics over $\operatorname{Spec} \mathbf{Z}$ that are reducible but nonsingular? Classify them.

The fibers over (2) and (5) are singular, however: modulo 2, we have

$$x^2 - y^2 - 5 = (x + y + 1)^2$$

and modulo 5 we can write

$$x^2 - y^2 - 5 = (x + y)(x - y)$$

Thus the fiber over (2) is a double line, while the fiber over (5) is a union of two lines (so that, in particular, there are two nonclosed points mapping to the point (5), while there is only one such point mapping to each of the other points $(p) \in \operatorname{Spec} \mathbf{Z}$).

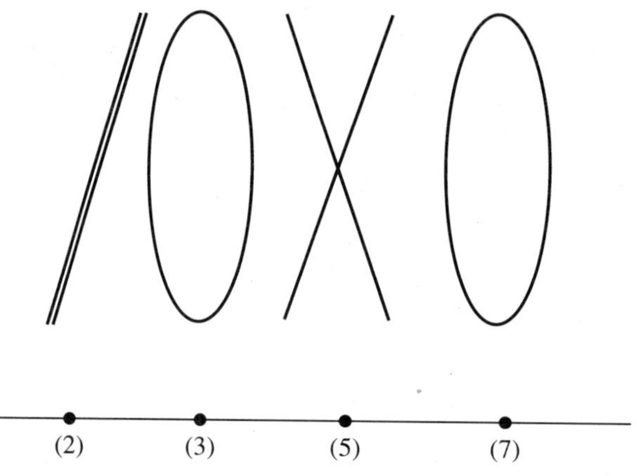

EXERCISE II-37

(This assumes some knowledge of projective geometry.) Note that the fiber of X over a point $(p) \in \operatorname{Spec} \mathbf{Z}$ such that $p \equiv 1 \bmod(4)$, $p \neq 5$, is really a hyperbola—that is, meeting the "line at infinity" in the fiber $\mathbf{A}^2_{\mathbf{F}_p}$ in two points with residue field $\mathbf{F}_p$, and isomorphic to $\mathbf{A}^1_{\mathbf{F}_p} - \{0\}$. Thus, for example, the fiber over (p) is the curve $x^2 + y^2 - 5 = 0$; its closure in the projective plane over $\mathbf{F}_p$ has equation $X^2 + Y^2 - 5Z^2 = 0$, and so meets the line $(Z = 0)$ at ∞ in the two points $[1, \alpha, 0]$ where $\alpha^2 \equiv p - 1 \bmod(p)$. Show that, by constrast, if $p \equiv 3 \bmod(4)$, the fiber is an ellipse; that is, it meets the line at ∞ in one point with residue field $\mathbf{F}_{p^2}$.

The preceding picture is very much in keeping with the geometric analogy: a surface fibered over a curve—such as, for example, the surface $x^2 - y^2 - z$ fibered over the z-line—will have a finite number of singular fibers, as in this classic picture:

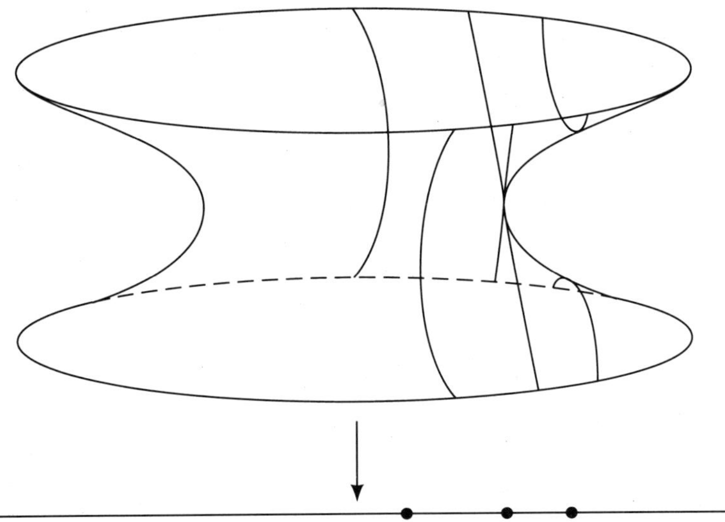

v Double Points in A_Z^1

Next, we consider some double points over $\mathbf{Z}$. Again, let

$$X = \mathbf{A}_\mathbf{Z}^1 = \operatorname{Spec} \mathbf{Z}[x]$$

If $Z \subset X$ is a closed subscheme supported at only one point, corresponding to a prime (p, f), say, we will wish to speak of the degree of Z just as we did in the case of finite subschemes over a field. In the case of schemes over a field, we defined the degree to be the dimension of $\mathcal{O}_Z(Z)$ as a vector space over k. But in the current case $\mathcal{O}_Z(Z)$ might contain no field at all—it might be $\mathbf{Z}/p^2$, for example. More confusing still, its residue field might not be $\mathbf{Z}/p$. In the case at hand the cheapest way out of this dilemma is to note that the cardinality card $\mathcal{O}_Z(Z)$ is always of the form p^d and take the degree to be d—this is obviously the vector space dimension if $\mathcal{O}_Z(Z)$ happens to be a $\mathbf{Z}/p$ vector space. (A more sophisticated approach is to define the degree of a reduced closed point first as the vector space dimension over $\mathbf{Z}/p$ and then define the degree of Z by multiplying the degree of the reduced point by the multiplicity of Z at this point.)

Consider the subschemes of degree 2 supported at the point $(7, x)$. These behave in a manner analogous to subschemes of degree 2 in the affine plane over a field. The ideal I of such a subscheme will always contain the square of the maximal ideal $\mathfrak{p} = (7, x)$ and so will be generated by $\mathfrak{p}^2$ together with one element of $\mathfrak{p}$: thus,

$$I = I_{\alpha,\beta} = (49, 7x, x^2, \alpha 7 + \beta x)$$

for some $\alpha, \beta \in \mathbf{Z}$ not both divisible by 7. Note that I will depend only on the congruence classes of α and β in $\mathbf{Z}/7$; and multiplying the pair (α, β) simultaneously by a unit in $\mathbf{Z}/7$ will not change I either. Thus for each point $[\alpha, \beta]$ of the projective line over the field of seven elements we get a double point supported at $\mathfrak{p}$.

EXERCISE II-38

Show that this correspondence is one to one.

The set of subschemes of degree 2 supported at $(7, x)$ may thus be identified with the projective line $\mathbf{P}^1$ over the field $\mathbf{F}_7$, much as the set of subschemes of $\mathbf{A}^2$ over a field k may be identified with the projective line over that field. (The identification in either case is actually with the projectivization of the *Zariski tangent space* to the ambient space at the point.) There is, however, one difference: whereas all subschemes of $\mathbf{A}_k^2$ of degree 2 supported at a point are isomorphic, the subschemes $Z_{\alpha,\beta}$ defined by $I_{\alpha,\beta}$ look different, even abstractly. We have

$$Z_{\alpha,\beta} = \operatorname{Spec} \mathbf{Z}/(49), \qquad \text{if } \beta \neq 0$$

but

$$Z_{\alpha,0} = \operatorname{Spec}(\mathbf{Z}/7)[x]/(x^2)$$

which are *not* isomorphic.

EXERCISE II-39

Classify (a) the subschemes of degree 3 supported at the point $(7, x)$, and (b) the subschemes of length 4 supported at the point $(2, x^2 + x + 1)$.

EXERCISE II-40

Referring to the diagram on page 77, use the preceding discussion to justify the fact that the curves $(4x - 5)$ and $(x - 2)$ are drawn tangent to one another, while the curves $(4x - 5)$ and (11) are drawn transverse.

Finally, here is an example of a flat family over Spec $\mathbf{Z}$. Recall that in the preceding section there was a discussion of the family of pairs of lines $M \cup L_t$, where M is the line $(x = z = 0)$ and L_t is the line $(y = z - t = 0)$. The key observation there was that the flat limit of the schemes $M \cup L_t$ as t approached zero was not the scheme $M \cup L_0$ but, rather, that scheme with an embedded point at the origin.

Here is the analogous phenomenon in a family parametrized by Spec $\mathbf{Z}$. Let $U = \operatorname{Spec} \mathbf{Z}[7^{-1}] = \operatorname{Spec} \mathbf{Z} - \{(7)\}$ be the complement of the point $(7) \in \operatorname{Spec} \mathbf{Z}$, and let

$$W = \mathbf{A}_U^3 := \operatorname{Spec} \mathbf{Z}[7^{-1}, x, y, z] \subset \mathbf{A}_{\mathbf{Z}}^3$$

be the corresponding open subscheme of $\mathbf{A}_{\mathbf{Z}}^3$. Let $\mathcal{N}$, $\mathcal{L}$ be the closed subschemes of $\mathbf{A}^3$ given by the ideals (x, z) and $(y, z - 7)$, respectively, and let $\tilde{\mathcal{N}} = \mathcal{N} \cap W$ and $\tilde{\mathcal{L}} = \mathcal{L} \cap W$. Let $\tilde{\mathcal{X}}$ be the union of $\tilde{\mathcal{N}}$ and $\tilde{\mathcal{L}}$, and let $\mathcal{X} \subset \mathbf{A}^3$ be the closure of $\tilde{\mathcal{X}}$ in $\mathbf{A}^3$. We may then think of $\tilde{\mathcal{X}}$ as a family of pairs of lines parametrized by U; and the fiber $\mathcal{X}_7$ of $\mathcal{X}$ over $(7) \in \operatorname{Spec} \mathbf{A}$ is the flat limit of this family "as 7 goes to 0." The fiber $\mathcal{X}_7$ is, as we expect, simply the union of the fibers $(x = z = 0)$ and $(y = z = 0)$ of $\mathcal{N}$ and $\mathcal{L}$ over (7); but the scheme $\mathcal{X}_7$ is not reduced: exactly as in the picture in the preceding section, it has an embedded point at the origin.

EXERCISE II-41

Verify the flatness of $\mathcal{X}$ over Spec $\mathbf{Z}$ and the description of $\mathcal{X}_7$. Can you find analogues over Spec $\mathbf{Z}$ of the other flat families discussed in the preceding section?

III

Projective Schemes

Once we have understood affine schemes, the theory of projective schemes does not really contain so much that is still novel: for the most part it differs from the classical theory of projective varieties in ways that are completely analogous to the difference between affine schemes and affine varieties. Nevertheless, there are some interesting examples of projective schemes that are not varieties and that do not really have a good affine analogue, and they arise in very natural contexts.

We start by describing the attributes, *separated* and *proper*, in scheme theory that correspond to the attributes of Hausdorffness and compactness in most of geometry. It is partly because projective varieties and schemes have these properties that they are fundamental objects in classical algebraic geometry and in the theory of schemes.

The next part of the chapter is devoted to the introduction of projective schemes and some examples. Just as in the case of affine schemes, two approaches to projective schemes are possible: one can define projective space and then take subschemes, or one can define all projective schemes on an equal footing, starting with graded algebras. As we did in the affine case, we adopt the second possibility. We give some examples, of which the most exotic are the projective double lines.

The final section of the chapter is devoted to three invariants of projective schemes embedded in projective space that were introduced by David Hilbert: the Hilbert polynomial, Hilbert function, and free resolution. Using these, we can sometimes distinguish among similar schemes, such as the projective double lines, and we can also shed some new light on the phenomenon of flatness.

A   Properness and Separation

Many techniques of geometry yield the most complete results when applied to compact Hausdorff spaces. Although affine schemes are quasicompact in the Zariski topology, they do not share the good properties of compact spaces in other theories because the Zariski topology is not Hausdorff. For example, the image of a regular map of affine schemes $\varphi : X \to Y$ need not be closed, even though X is quasicompact.

Moreover, the fact that the Zariski topology is not Hausdorff has another unpleasant consequence. Recall that in the general definition of a manifold, one starts with a topological space that is Hausdorff and admits a covering by charts of the standard form (balls in Euclidean space, say). The fact that the balls themselves are Hausdorff is not enough by itself to guarantee that the total space is. This is why the line with the doubled origin described in Exercise I-41, and shown here,

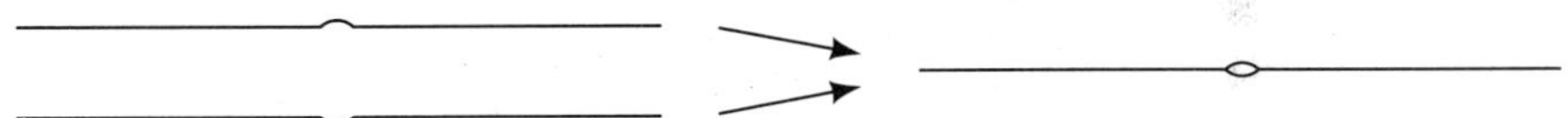

is not a manifold. However, when we work with schemes (or, for that matter, with varieties) glued together from affine schemes, we cannot afford to specify that the total space is Hausdorff, because even the local pieces are not. This has the result that given two maps of schemes $\varphi, \psi : X \to Y$, the set where φ and ψ are equal may not be closed, as in the following exercise, which is a typical case.

EXERCISE III-1 ____

a. Let Y be the line with doubled origin over a field k, defined in Exercise I-41, and let $\varphi_1, \varphi_2 : \mathbf{A}^1 \to Y$ be the two obvious inclusions. Show that the locus where φ_1 and φ_2 agree (simply as continuous maps of topological spaces) is not closed.

b. Now let $X = Y \times_k Y$, and let φ and ψ be the two projection maps from X to Y. Show that the set of points at which φ and ψ agree is not closed (note that this is just the diagonal, defined below). Show that the same is true for the set of closed points at which φ and ψ agree, so this is not a pathology special to schemes but occurs already in the category of varieties.

Such a pathology cannot happen, however, if X is an affine scheme; nor, it turns out, can it happen when X is a projective scheme. The desirable property that these schemes have, which is one of the most important consequences of the Hausdorff property for manifolds, is expressed by saying that X is *separated* as a scheme over k. In general, given any map $\alpha : X \to S$ of schemes, we define the *diagonal* subscheme $\Delta \subset X \times_S X$ to be the subscheme defined locally on $X \times X$ for each affine open

$$X \supset \operatorname{Spec} A \xrightarrow{\alpha|_{\operatorname{Spec} A}} \operatorname{Spec} B \subset S$$

by the ideal I generated by all elements of the form

$$a \otimes 1 - 1 \otimes a \in A \otimes_B A$$

We then say that α is separated, or that X is separated as a scheme over S, if $\Delta \subset X \times_S X$ is a closed subscheme.

EXERCISE III-2

Let $X \to S$ be any map of topological spaces, and let

$$\Delta \subset X \times_S X$$

be the diagonal. Show that if Δ is a closed set, then for any commutative diagram of continuous functions,

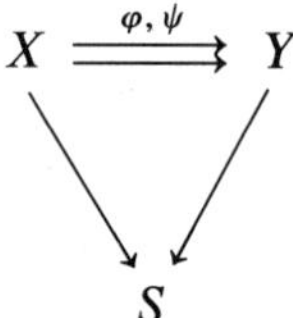

the set of points of X where φ and ψ agree is closed. Now prove a similar lemma for regular maps of schemes: show that there is a naturally defined (that is, maximal) closed subscheme on which φ and ψ agree.

EXERCISE III-3

Let X be a scheme separated over S. Show that (closed or open) subschemes of X are again separated over S.

EXERCISE III-4

Note that from the very definition of the diagonal it follows that affine schemes are separated.

We shall see below that projective schemes, to be defined shortly, are also separated, so at least these features of the properties of Hausdorff spaces are valid for them as well.

In the case of classical affine varieties—even things as simple as plane curves—it was realized early in the previous century that the simplest way to get something that would behave like a compact object—would, in fact, be compact in the "classical" topology, in the case of varieties over the complex numbers—was to take the closure of an affine variety in projective space. It turns out that if $\varphi : X \to Y$ is a map of projective varieties, then indeed φ maps closed subvarieties of X to closed subvarieties of Y. Somewhat more generally, if we take the product of such a map with an arbitrary variety Z, to get

$$\psi := \varphi \times 1_Z : X \times Z \to Y \times Z$$

then ψ maps closed subvarieties of $X \times Z$ onto closed subvarieties of $Y \times Z$. It turns out that *this, with the separation property, is the central property of projective varieties that makes them so useful.* But it is a property satisfied by a slightly larger class of varieties than the projective ones, and it is a property that is sometimes easier to verify than projectivity, so it is of great importance to make a general definition.

If $\alpha : X \to S$ is a map of schemes, then we will say that α is *proper*, or that X is *proper over S*, if α is separated and for all maps $T \to S$, the projection map of the fibered product

$$X \times_S T \to T$$

carries closed subsets onto closed subsets. As usual, if $S = \operatorname{Spec} R$ is a ring, we shall often say "proper over R" when we mean "proper over $\operatorname{Spec} R$."

The additional property given here, besides that of separation, is sometimes expressed by saying that α is *universally closed*. The name *proper* comes from an old geometric usage: a map $\alpha : M \to N$ of Hausdorff spaces is called proper if the preimage of every compact set is compact. This is a kind of "relative" compactness for the map α. It is related to our notion by the property expressed in the following exercise.

EXERCISE III-5 ____

Show that if $M \to N$ is a map of Hausdorff topological spaces, then it is proper in the old sense (in the category of Hausdorff spaces and continuous maps) if and only if it is universally closed in this category.

This notion of properness turns out to be the key property in algebraic geometry—whether of schemes or of varieties—that plays the role played by "compact and Hausdorff" in other geometric theories. The projective schemes (over a ring S) that we will introduce below are simply the simplest and (therefore?) most important examples of varieties proper over S. We will not prove this central result here; it is not terribly difficult, but it would take us too far afield. See, for example, Hartshorne (1977, Chap. II) for a proof.

B   Proj of a Graded Ring

i   The Construction of Proj S

By far the most important examples of schemes that are not affine are the schemes *projective* over an affine scheme $\operatorname{Spec} k$, where k is an arbitrary commutative ring. (For simplicity we usually say that such a scheme is projective over k instead of over $\operatorname{Spec} k$.) Such a scheme is obtained from a graded k-algebra by a process very much analogous to the construction of a projective variety from its homogeneous coordinate ring. One can also define schemes projective over an arbitrary base scheme B by starting with a sheaf of graded $\mathcal{O}_B$-algebras, and this generalization has important applications. But most of the theory quickly reduces to the case where B is affine, and we will stick with that level of generality here.

To describe this construction, we start with a finitely generated, positively graded k-algebra having k as the degree 0 part—that is, a finitely generated k-algebra S with a *grading*

$$S = \bigoplus_{v=0}^{\infty} S_v \qquad (\text{as } k\text{-modules})$$

such that

$$S_v S_\mu \subset S_{v+\mu}$$

and

$$S_0 = k$$

A element of S is called *homogeneous* of degree v if it lies in S_v. We will define a k-scheme $X = \text{Proj}\, S$ from S. The schemes *projective over* k are by definition the schemes of the form $\text{Proj}\, S$. The algebra S is called the *homogeneous coordinate ring* of X, though (like the homogeneous coordinate ring of a projective variety) it is in fact *not* determined by X.

In case S is the polynomial ring

$$S = k[x_0, \ldots, x_r]$$

over k, with grading defined by giving the elements of k degree 0 and giving each variable degree 1, the resulting scheme $\text{Proj}\, S$ is called projective r-space over k and is usually written $\mathbf{P}_k^r$. In case k is a field, the scheme $\mathbf{P}_k^r$ bears the same relation to the variety called projective space over k as the scheme $\mathbf{A}_k^r$ bears to the variety called affine r-space. As usual, we shall drop the subscript k when it is clear from context or irrelevant.

We will suppose for simplicity that, as in the case of the polynomial ring, the algebra S is generated over k by its elements of degree 1, and we leave the general case as an exercise. (In a different direction, most of what we say below also holds if S is not assumed to be finitely generated over k, but this generalization is less frequently used.)

$\text{Proj}\, S$ may be defined as follows: we write $S_+ = \bigoplus_{v=1}^{\infty} S_v$ for the ideal generated by homogeneous elements of strictly positive degree in S. We say that an ideal is homogeneous if it is generated by homogeneous elements. The underlying topological space $|\text{Proj}\, S|$ is the set of homogeneous prime ideals in the ring S that do not contain S_+ (these are sometimes called *relevant* prime ideals, and S_+ is thus called the *irrelevant ideal*). The topology of $|\text{Proj}\, S|$ is defined by taking the closed sets to be the sets of the form

$$Z_+(I) := \{\mathfrak{p} \,|\, \mathfrak{p} \text{ is a relevant prime of } S \text{ and } \mathfrak{p} \supset I\}$$

for some homogeneous ideal I of S.

We will give $\text{Proj}\, S$ the structure of a scheme by specifying this structure on each of a basis of open sets. To do this, let f be any homogeneous element of S of degree 1, and let U be the open set

$$|\text{Proj}\, S| - Z_+((f))$$

the set of homogeneous primes of S not containing f (and thus not containing S_+). The points of U may be identified with the homogeneous primes of $S[f^{-1}]$. On the other hand, these homogeneous primes correspond to all the primes of the ring of elements of degree 0 in $S[f^{-1}]$, which is denoted by $S[f^{-1}]_0$ (see part a of Exercise III-6). Thus we may identify U with the topological space $\operatorname{Spec} S[f^{-1}]_0$ and give it the corresponding structure of an affine scheme. We will write $(\operatorname{Proj} S)_f$ for this open affine subscheme of $\operatorname{Proj} S$. If $x_0, \ldots, x_r$ are elements of degree 1 generating an ideal whose radical is the irrelevant ideal S_+, then the open sets $(\operatorname{Proj} S)_{x_i} := \operatorname{Proj} S - Z_+(x_i)$ form an affine open cover of $\operatorname{Proj} S$.

If g is another degree 1 element of S, then the overlap $(\operatorname{Proj} S)_f \cap (\operatorname{Proj} S)_g$ is the open affine subset of $(\operatorname{Proj} S)_f$ given by the spectrum of $S[f^{-1}]_0[(g/f)^{-1}] = S[f^{-1}, g^{-1}]_0$. Since this expression is symmetric in f and g, we get a natural identification

$$((\operatorname{Proj} S)_f)_{(g/f)} = ((\operatorname{Proj} S)_g)_{(f/g)}$$

As in the discussion of gluing in Section I-B-ii-a, this makes $\operatorname{Proj} S$ into a scheme.

In the rest of this section and the next we present some basic facts about projective schemes and their closed subschemes. Since these facts and their proofs are quite parallel to things from the theory of varieties, we present them as exercises.

EXERCISE III-6 ───

Prove the following:

a. For any homogeneous ideal I of S and homogeneous element f of degree 1, the intersection

$$(I \cdot S[f^{-1}]) \cap S[f^{-1}]_0$$

is generated by elements obtained by choosing a set of homogeneous generators of I and multiplying them by the appropriate (negative) powers of f (see Exercise III-10 for the generalization where f has arbitrary degree). Thus the homogeneous primes of $S[f^{-1}]$ are in one-to-one correspondence with all the primes (no homogeneity condition) of the ring of elements of degree 0 in $S[f^{-1}]$; the correspondence is given by taking a prime $\mathfrak{p}$ of $S[f^{-1}]$ to $\mathfrak{q} = \mathfrak{p} \cap S[f^{-1}]_0$ and taking the prime $\mathfrak{q}$ of $S[f^{-1}]_0$ to $\mathfrak{q}S[f^{-1}]$.

b. Let $S = k[x_0, \ldots, x_r]$ be the polynomial ring, and let U be the open affine set $(\mathbf{P}_k^r)_{x_i}$ of $\mathbf{P}_k^r = \operatorname{Proj} S$. By definition,

$$U = \operatorname{Spec} S[x_i^{-1}]_0$$

Show that

$$S[x_i^{-1}]_0 = k[x_0', \ldots, x_r']$$

the polynomial ring with generators $x_j' = x_j/x_i$. (Note that $x_i' = 1$, so that this is a polynomial ring in r variables.) Thus

$$(\mathbf{P}_k^r)_{x_i} = \mathbf{A}_k^r$$

so projective r-space has an open affine cover by $r + 1$ copies of affine r-space, as one would expect from a knowledge of projective space as an algebraic variety over an algebraically closed field, or as a complex manifold.

Consider the map $\alpha : S \to S[x_i^{-1}]_0$ obtained by mapping x_i to 1 and x_j to x_j' for $j \neq i$. Show from part a that if I is a homogeneous ideal of S then $I' := I \cdot S[x_i^{-1}] \cap S[x_i^{-1}]_0 = \alpha(I) \cdot S[x_i^{-1}]_0$. The process of making I' from I is called *dehomogenization*. Describe, as in the classical case, the inverse process, homogenization.

EXERCISE III-7

If I is a homogeneous ideal of the graded ring S, then we have an inclusion of underlying sets

$$|\operatorname{Proj} S/I| \subset |\operatorname{Proj} S|$$

Show that the intersection of this subset with an open affine $(\operatorname{Proj} S)_f$ is a closed subset of $(\operatorname{Proj} S)_f$, and that the corresponding subscheme is isomorphic to $(\operatorname{Proj} S/I)_f$, so that $\operatorname{Proj} S/I$ can be realized as a closed subscheme of $\operatorname{Proj} S$. Every finitely generated k-algebra generated in degree 1 is a factor ring, by a homogeneous ideal, of the polynomial ring $k[x_0, \ldots, x_r]$ for some r, so we see that *every projective scheme over k is a closed subscheme of a projective space over k*. We will see in more detail the correspondence between ideals in the ring S and closed subschemes of $\operatorname{Proj} S$ in Exercises III-22 and III-23.

EXERCISE III-8

Show that $\mathbf{P}_k^r$ is the disjoint union of the open set $\mathbf{A}_k^r$ and the closed set $\mathbf{P}_k^{r-1}$. In particular, $\mathbf{P}_k^0 = \operatorname{Spec} k$. Thus, for example, we may picture $\mathbf{P}_\mathbf{Z}^1$ as the union of the affine line $\mathbf{A}_\mathbf{Z}^1$ over $\mathbf{Z}$ (as pictured in Chapter II) with a "point at ∞" isomorphic to $\operatorname{Spec} \mathbf{Z}$, as follows:

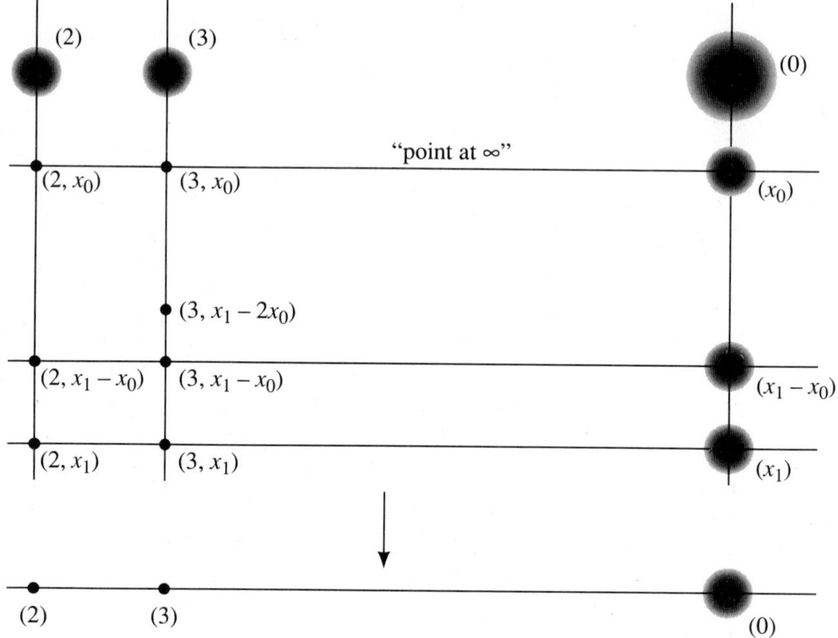

EXERCISE III-9

Add to this diagram pictures of the closures of the points $(4x_1 - 5x_0)$, $(2x_1 - 5x_0)$, and (5) (compare with the diagram of $\mathbf{A}_{\mathbf{Z}}^2$ in Section II-D-iii). *Note:* The curve $(4x_1 - 5x_0)$ should be drawn tangent to the "point at ∞" (x_0), while the curve $(2x_1 - 5x_0)$ should not—informally, we could say this is because "the function $5/4$ has a double pole at (2), while $5/2$ has only a simple pole there." (See also the discussion in the beginning of Section II-D-v.)

EXERCISE III-10

With notation as above, let h be a homogeneous element of S of any strictly positive degree. The set

$$(\mathrm{Proj}\, S)_h := (\mathrm{Proj}\, S) - Z_+((h))$$

is as above the set of homogeneous primes of S not containing h. Show that this set is again in one-to-one correspondence with the set of primes of $S[h^{-1}]_0$ and that in fact there is an isomorphism of $\mathrm{Spec}\, S[h^{-1}]_0$ with an open (affine) subscheme of $\mathrm{Proj}\, S$. Show also that a collection

$$\{(\mathrm{Proj}\, S)_h\}_{h \in H}$$

of such open affines is an open cover of Proj S iff the elements of H generate an ideal whose radical equals S_+.

EXERCISE III-11*

Extend the definition of Proj S to the case where S is not necessarily generated by elements of degree 1, and show that Proj S is a projective scheme.

EXERCISE III-12

Let S be a graded ring, not necessarily generated in degree 1. For any positive integer d, define the dth *Veronese subring* of S to be the graded ring

$$S^{(d)} = \bigoplus_{v=0}^{\infty} S_{dv}$$

Show that Proj S is isomorphic to Proj $S^{(d)}$. However, show that if $S = k[x, y]$, then $S^{(d)}$ is not isomorphic to S as a graded algebra (or even as a ring). Thus, as in the case of varieties, the correspondence between graded algebras and projective schemes is not one to one.

ii Morphisms to Projective Space and Invertible Sheaves

Just as there is a simple characterization of morphisms to an affine scheme (Theorem I-36), there is a simple way of viewing morphisms to projective space in terms of line bundles, or, equivalently, invertible sheaves, and we will derive it in this section. There is a third kind of object that may also be used in the description: linear series of Cartier divisors. But for this we refer the reader to one of the standard texts, such as Hartshorne's.

If we understand morphisms to the scheme $\mathbf{P}_k^n$, we will understand morphisms to an arbitrary projective scheme $Y \subset \mathbf{P}_k^n$, since a morphism to Y is just a morphism to $\mathbf{P}_k^n$ that factors through Y (a sharp version of this is given in Exercise III-26); thus we will study morphisms to projective space.

Looking to other geometric theories for a hint, we recall that in topology the space $\mathbf{P}_{\mathbf{C}}^n$ of n-dimensional subspaces of a complex $n + 1$-dimensional vector space $\mathbf{C}^{n+1}$ is the classifying space for subbundles of rank n of a trivial bundle of rank $n + 1$ (and similarly for $\mathbf{P}_{\mathbf{R}}^n$). This means that for all spaces X, maps $X \to \mathbf{P}_{\mathbf{C}}^n$ correspond to the rank n subbundles of the trivial bundle on X. The correspondence is easy to describe: a rank n subbundle $\mathscr{S}$ of the trivial bundle $\mathscr{V} = \mathbf{C}^n \times X$ on X corresponds to the map $X \to \mathbf{P}_{\mathbf{C}}^n$ that sends a point $p \in X$ to the point of $\mathbf{P}_{\mathbf{C}}^n$ corresponding to the space

$$\mathscr{S}_p \subset \mathscr{V}_p = \mathbf{C}^n \times \{p\} = \mathbf{C}^n$$

There are other equivalent descriptions, which may be more familiar, in terms of the rank 1 quotient bundle $\mathscr{V}/\mathscr{S}$, or the rank 1 subbundle $(\mathscr{V}/\mathscr{S})^* \subset \mathscr{V}^*$.

Analogous results hold in the category of complex analytic spaces and maps and in the category of algebraic varieties and regular maps (taking the subbundles to be complex analytic, or algebraic, respectively).

In this section we want to give a corresponding result for schemes. The main difference is that in algebraic geometry, it is traditional to replace vector bundles on X by their sheaves of sections.

To see what these sheaves should look like, consider first that if $\mathscr{E}$ is a trivial vector bundle of rank 1 on a scheme X, then a section of $\mathscr{E}$ is the same as a function on X, so the sheaf of sections of $\mathscr{E}$ should be $\mathscr{O}_X$. Taking direct sums, we see that the sheaf of sections of a trivial vector bundle of rank m is the coherent sheaf that is the free $\mathscr{O}_X$-module $\mathscr{O}_X^m$. In general, since vector bundles are by definition locally trivial, their sheaves of sections are locally free sheaves of $\mathscr{O}_X$-modules of finite rank—locally free coherent sheaves. It is not hard to go in the other direction as well and to derive from a locally free coherent sheaf a vector bundle.

Given this equivalence between vector bundles and locally free coherent sheaves, why work with locally free sheaves? The reason is similar to the reason for working with schemes instead of varieties even if one is primarily interested in varieties: locally free coherent sheaves live naturally in the larger category of coherent sheaves, and working in the larger category gives us flexibility. Standard constructions in the smaller category (such as taking the fibers of a morphism of schemes, or taking the cokernel of a homomorphism of locally free sheaves) are most naturally interpreted in the large category.

Like the line bundles to which they correspond, locally free sheaves of rank 1 play an especially important role and have a special name: they are called *invertible sheaves*. (The terminology comes from number theory: an invertible module over a domain T is a finitely generated submodule I of the quotient field such that, for some other finitely generated submodule J of the quotient field (called its inverse) we have $IJ = T$, the unit ideal. Over the scheme $\mathrm{Spec}(T)$, the corresponding sheaf is an invertible sheaf.)

We are now ready to characterize morphisms from a scheme to a projective space.

THEOREM III-13   If X is a k-scheme, then

$$\mathrm{Morph}_{k-\mathrm{schemes}}(X, \mathbf{P}_k^n)$$

$$= \{\text{subsheaves } K \subset \mathscr{O}_X^{n+1} \text{ that locally are summands of rank } n\}$$

$$= \frac{\{\text{invertible sheaves } P \text{ on } X \text{ with an epimorphism } \mathscr{O}_X^{n+1} \to P\}}{\{\text{units of } \mathscr{O}_X(X) \text{ acting as automorphisms of } P\}}$$

Because all the terms in these equalities are defined locally on X, the theorem reduces easily to the case where X is affine, and this is the case we will actually prove below. First, we review the corresponding notions about modules. A good basic reference is Bourbaki (1989, Chap. II-5).

Recall that a module K over a ring T is *locally free* of rank m if for every maximal ideal (or, equivalently, every prime ideal) $\mathfrak{p}$ the $T_{\mathfrak{p}}$-module $K_{\mathfrak{p}}$ is free of rank m. This is the same as the sheaf-theoretic notion.

EXERCISE III-14 ____

Let K be a finitely generated module over a ring T, and let $\tilde{K}$ be the corresponding coherent sheaf over $\operatorname{Spec}(T)$. Show that K is a locally free module in the sense above iff $\tilde{K}$ is a locally free coherent sheaf in the sense that there is an affine cover of $\operatorname{Spec}(T)$ by basic open sets $\operatorname{Spec}(T)_{f_i}$ such that the restriction of $\tilde{K}$ to each of these sets is free (equivalently, each $K[f_i^{-1}]$ is free over $T[f_i^{-1}]$.)

An *invertible* T-module is a finitely generated, locally free T-module of rank 1.

In commutative algebra, locally free modules are usually called projective modules; their characteristic property is that if P is a locally free T-module, then any epimorphism of T-modules $M \twoheadrightarrow P$ splits. It follows that if $K \subset T^{n+1}$ is a submodule, then K is a summand of T^{n+1} iff T^{n+1}/K is a locally free module; in particular, K is a rank n summand of T^{n+1} iff T^{n+1}/K is an invertible module.

Before giving the proof of Theorem III-13, we record a result that comes from an immediate application of the definitions.

PROPOSITION III-15   A morphism of an arbitrary scheme X to a projective space $\mathbf{P}_k^r = \operatorname{Proj} k[x_0, \ldots, x_r]$ may be given by a collection of maps $U_i \to (\mathbf{P}_k^r)_{x_i}$, where $\{U_i\}$ is an open cover of X, the $(\mathbf{P}_k^r)_{x_i} \subset \mathbf{P}_k^r$ are the open subsets of Exercise III-6, and the maps φ_i and φ_j induce the same map $U_i \cap U_j \to (\mathbf{P}_k^r)_{x_i} \cap (\mathbf{P}_k^r)_{x_j} = \operatorname{Spec}(k[x_0, \ldots, x_r][x_i^{-1}, x_j^{-1}]_0)$. $\square$

The heart of Theorem III-13 is the following result, which is the affine version of the first equality.

PROPOSITION III-16   If T is a Noetherian k-algebra, then

$$\operatorname{Morph}_{k-\text{schemes}}(\operatorname{Spec}(T), \mathbf{P}_k^n)$$

$$= \{K \subset T^{n+1} \mid K \text{ is locally a rank } n \text{ direct summand of } T^{n+1}\}$$

Proof   Suppose, first, that K is a rank n free summand of T^{n+1}, and write P for the module T^{n+1}/K. This module is locally free of rank 1 and is generated by the $n+1$ images e_i of the $n+1$ generators of T^{n+1}. Let $I_j = $ annihilator (P/Te_j), and let U_j be the complement of $V(I_j)$ in Spec T, so that the U_j form an open cover of Spec T. Regard T-modules as sheaves on Spec T. On U_j the map $T \to P$ defined by $1 \mapsto e_j$ is an isomorphism, and identifying $P|_{U_j}$ with $T|_{U_j}$ via this map, the projection $T^{n+1}|_{U_j} \to (T^{n+1}/K)|_{U_j} = P|_{U_j} = T|_{U_j}$ has a matrix of the form $(t_{j0}, \ldots, t_{jj} = 1, \ldots, t_{jn})$, which defines an element of $T^{n+1}|_{U_j}$ and thus a morphism $\mathrm{Spec}(T) \to \mathbf{A}_k^n$. These morphisms agree on overlaps as in III-15, so they define a morphism Spec $T \to \mathbf{P}_k^n$.

Conversely, suppose that we are given a morphism ψ from Spec T to $\mathbf{P}_k^n$. Since $\mathbf{P}_k^n$ is covered by $n+1$ affine n-spaces, ψ is by definition associated to an open cover Spec $T = \bigcup_{j=0,\ldots,n} U_j$, and for each j an element $(t_{j0}, \ldots, t_{jj} = 1, \ldots, t_{jn})$ of $T^{n+1}|_{U_j}$, such that t_{ij} is a unit on $U_i \cap U_j$ and $t_{i\ell} = t_{ij}t_{j\ell}$ in $T|_{U_i \cap U_j}$ for all i, j, ℓ. Two such T-valued points are the same iff the corresponding elements of $T^{n+1}|_{U_j}$ are equal for each j. Let K_i be the kernel of the map

$$T^{n+1}|_{U_j} \to T|_{U_j}$$

defined by the matrix $(t_{j0}, \ldots, t_{jj} = 1, \ldots, t_{jn})$, and let

$$K = \{a \in T^{n+1} \mid a|_{U_j} \in K_j \text{ for each } j\}$$

To see that K is locally a rank n summand of T^{n+1}, note that any local ring of T is a local ring of one of the U_j, so the result of localizing the sequence

$$0 \to K \to T^{n+1} \to T^{n+1}/K \to 0$$

at any prime ideal $\mathfrak{p}$ is a sequence of the form

$$0 \to K_\mathfrak{p} \to T_\mathfrak{p}^{n+1} \to T_\mathfrak{p} \to 0.$$

Such a sequence must split. $\square$

<hr>

EXERCISE III-17

If T is Noetherian, then the word *locally* can be omitted in the statement of the proposition. This is because in the Noetherian case a submodule of a finitely generated module that is locally a direct summand is in fact a direct summand. Prove this.

To derive a version of this with invertible modules, we use the fact that $K \subset T^{n+1}$ is a direct summand of rank n iff T^{n+1}/K is an invertible module, and identify the set on the right-hand side of the equality in III-16 with the set of invertible quotient modules of T^{n+1}. We may separate the isomorphism class of the quotient from the surjection that makes it a quotient and look at invertible T-modules P with surjections $T^{n+1} \to P$. Two surjections α, $\beta : T^{n+1} \to P$ have the same kernel iff there is an automorphism $\sigma : P \to P$ such that $\beta = \sigma\alpha$. But if P is an invertible T-module, then $\mathrm{Hom}_T(P, P) = T$ (reason: the natural map $\alpha : T \to \mathrm{Hom}_T(P, P)$ taking 1 to the identity is locally the same as the natural map $T \to \mathrm{Hom}_T(T, T)$, which is an isomorphism, so α is an isomorphism). Thus the automorphisms of P may be identified with units of T, and we get the following corollary.

COROLLARY III-18   If T is a k-algebra, then

$$\mathrm{Morph}_{k-\mathrm{schemes}}(\mathrm{Spec}(T), \mathbf{P}_k^n)$$

$$= \frac{\{\text{invertible } T\text{-modules } P \text{ with an epimorphism } T^{n+1} \to P\}}{\{\text{units of } T \text{ acting as automorphisms of } P\}}$$

In case T is a local ring, all this is particularly simple. In the first place, any invertible T-module is isomorphic to T itself. Thus a homomorphism $T^{n+1} \to T$ corresponds to an $(n + 1)$-tuple of elements of k, and the homomorphism is an epimorphism iff one of the elements is a unit. Thus we get the next corollary.

COROLLARY III-19   If T is a local k-algebra, then

$$\mathrm{Morph}_{k-\mathrm{schemes}}(\mathrm{Spec}(T), \mathbf{P}_k^n)$$

$$= \frac{\{n + 1\text{-tuples of elements of } T, \text{ at least one a unit}\}}{\{\text{units of } T \text{ acting by multiplication}\}}$$

Note that in the case where $T = k$ this is the familiar statement that the (closed) points of $\mathbf{P}_k^n$ are just $(n + 1)$-tuples of elements of k, modulo nonzero elements of k!

iii Closed Subschemes of Proj R

As might be expected, closed subschemes of $\mathbf{P}_k^r$ come from relevant homogeneous ideals of the ring $k[x_0, \ldots, x_r]$ in the sense that a homogeneous ideal

$I \subset k[x_0,\ldots,x_r]$ does determine a coherent sheaf of ideals $\tilde{I} \subset \mathcal{O}_{\mathbf{P}^r}$. The following problems deal with the necessary facts and definitions.

EXERCISE III-20

For each open set

$$U_i = (\mathbf{P}^r_k)_{x_i} = \operatorname{Spec} k[x_0,\ldots,x_r,x_i^{-1}]_0 \cong \mathbf{A}^r_k$$

let $\tilde{I}(U_i)$ be the ideal $I \cdot k[x_0,\ldots,x_r,x_i^{-1}] \cap k[x_0,\ldots,x_r,x_i^{-1}]_0$. Show that this definition may be extended in a unique way to other open sets U in such a way that $\tilde{I}$ becomes a coherent sheaf of ideals. We may thus speak of the *closed subscheme $V(\tilde{I})$ of $\mathbf{P}^r_k$ associated to a homogeneous ideal I.*

EXERCISE III-21

Conversely, given a closed subscheme X in $\mathbf{P}^r_k$, we may define a homogeneous ideal $I(X) \subset k[x_0,\ldots,x_r]$ to be the ideal generated by all homogeneous polynomials $p(x_0,\ldots,x_r)$ such that for every i setting the ith variable equal to 1 gives rise to an element

$$p(x_0,\ldots,1,\ldots,x_r) \in \mathscr{I}_X(U_i) \subset k[x_0,\ldots,x_r,x_i^{-1}]_0$$

Show that if $I = I(X)$, then $\tilde{I} = \mathscr{I}_X$.

Note that with Exercise III-7 this shows that every closed subscheme of a projective scheme is projective: if $I \subset S = k[x_0,\ldots,x_r]$ is a homogeneous ideal, then $V(\tilde{I}) \subset \mathbf{P}^r_k$ is isomorphic to the scheme $\operatorname{Proj} S/I$.

EXERCISE III-22

The correspondence between subschemes and ideals is not, as it was in the case of affine schemes, one to one. For example, show that in $\mathbf{P}^1_k$, with k a field, the ideals $I = (x_0)$ and $I' = (x_0^2, x_0 x_1)$ both define the same reduced, one-point subscheme. More generally, show that if $I \subset S = k[x_1,\ldots,x_r]$ is any homogeneous ideal, and for any integer n_0 we define an ideal $I' \subset I$ by

$$I' = \bigoplus_{n \geq n_0} I_n$$

then I and I' define the same subscheme of $\mathbf{P}^r_k$.

EXERCISE III-23

To deal with this, define the *saturation* of a homogeneous ideal $J \subset S :=
k[x_0, \ldots, x_r]$ to be the ideal

$$I = \{F \in S : \text{for some } n,\ F \cdot S_n \subset I\}$$

and say that a homogeneous ideal is saturated if it equals its saturation. Show
that there is a bijective correspondence between subschemes of $\mathbf{P}_k^r$ and saturated ideals.

EXERCISE III-24

Show that the isomorphism of Exercise III-12 defines an isomorphism
between projective space $\mathbf{P}_k^r$ and a closed subscheme of $\mathbf{P}_k^{N-1}$, where
$N = \dim_k(k[x_0, \ldots, x_r]_d)$ (this is just the scheme-theoretic version of the
Veronese map).

EXERCISE III-25

Show that if R is any graded ring finitely generated over a ring $k = R_0$—but
not necessarily generated by its graded part of degree 1—then $\operatorname{Proj} R$ is
isomorphic to a closed subscheme of some projective space $\mathbf{P}_k^r$.

EXERCISE III-26

   a. Suppose that $Y \subset \mathbf{P}_k^n$ is the closed subscheme defined by homogeneous equations $\{F_i\}$. If T is a local k-algebra, then as we showed above, the
morphisms from $\operatorname{Spec}(T)$ to $\mathbf{P}_k^n$ may be identified with $n + 1$-tuples of elements of T generating the unit ideal, modulo units of T. Show that the
condition that such an $n + 1$-tuple correspond to a map to Y is simply that
it be a zero of all the polynomials F_i.

   b. The general case of a map from an affine k-scheme to a projective
k-scheme can be reduced to the local one using the following fact: if T is any
k-algebra a morphism $\operatorname{Spec} T \to \mathbf{P}_k^n$ factors through X iff for all primes $\mathfrak{p}$ of T
the composite morphisms $\operatorname{Spec} T_{\mathfrak{p}} \to \operatorname{Spec} T \to \mathbf{P}_k^n$ factor through X. Prove
this.

iv Examples: Projective Double Lines

So far, everything we have seen has been parallel to the theory of varieties.
We will now look at one genuinely nonclassical family of examples.

At the end of Chapter II we asked the reader to show that all affine double lines are equivalent. This is not true for projective double lines. Here are some simple examples.

Let k be a field. Consider the graded ring

$$S := k[u, v, x, y]/(x^2, xy, y^2, u^d x - v^d y)$$

with all variables having degree 1, and the scheme

$$X := \operatorname{Proj} S$$

We claim that X is a double line (that is, it has an affine open cover, where each affine piece is an affine double line) and that the isomorphism class of X depends on the number d.

First of all, we construct an open affine covering of X. The elements x and y are nilpotent in S, so the radical of the ideal generated by u and v is the irrelevant ideal of S, and X is covered by X_u and X_v. From the definitions we see that

$$X_u = \operatorname{Spec}(S[u^{-1}])_0$$

To analyze the ring $(S[u^{-1}])_0$, we note that it is a factor ring of

$$k[u, v, x, y][u^{-1}]_0 = k[v', x', y']$$

where

$$v' = \frac{v}{u} \qquad x' = \frac{x}{u} \qquad y' = \frac{y}{u}$$

and the kernel of the map to $(S[u^{-1}])_0$ is generated by $(x')^2$, $x'y'$, $(y')^2$, and $x' - (v')^d y'$ (see Exercise III-6). Thus

$$(S[u^{-1}])_0 \cong k[v', y']/(y')^2$$

and X_u is an affine double line. By symmetry, X_v is too, and this proves that X is a projective double line. Explicitly,

$$X_v \cong \operatorname{Spec}(S[v^{-1}])_0$$

and

$$(S[v^{-1}])_0 \cong k[u'', x'', y'']/((x'')^2, x''y'', (y'')^2, (u'')^d x'' - y'')$$

$$\cong k[u'', x'']/(x'')^2$$

where

$$u'' = \frac{u}{v} = \frac{1}{v'} \qquad x'' = \frac{x}{v} \qquad y'' = \frac{y}{v}$$

As we indicated, in contrast to the affine case, not all double lines are isomorphic to one another. The simplest way to see this is to show that the isomorphism class of X depends on the integer d, which may be thought of as specifying how fast the double line twists around the reduced line inside it. To demonstrate this, we will show that the ring of global sections $\mathcal{O}_X(X)$ of the structure sheaf of X depends on d. To compute it, suppose first that $\sigma \in \mathcal{O}_X(X)$. The element σ restricts to an element of $\mathcal{O}_X(X_u)$, which is isomorphic to $k[v', y']/(y')^2$ by the above, so we may write

$$\sigma|_{X_u} = a(v') + b(v')y'$$

and similarly

$$\sigma|_{X_v} = f(u'') + g(u'')x''$$

for unique polynomials a, b, f, and g with coefficients in k. But on $X_u \cap X_v$ we have

$$u'' = \frac{1}{v'}$$

and

$$x'' = \frac{x}{v} = x'\left(\frac{u}{v}\right) = (v')^d y'\left(\frac{u}{v}\right) = (v')^{d-1}y'$$

Thus $f(1/v') = a(v')$, which is only possible if f and a are constant polynomials and $f = a$. Also, $g(1/v')(v')^{d-1} = b(v')$, which is only possible if both g and b have degree less than or equal to $d - 1$ (and then each of g and b determines the other). Conversely, any element of $\mathcal{O}_X(X_u)$ of the form $a + b(v')y'$ with a a constant and b a polynomial of degree less than or equal to $d - 1$ extends uniquely to a global section of $\mathcal{O}_X$, so we see that the dimension of $\mathcal{O}_X(X)$ is $d + 1$. This shows that the isomorphism class of X depends on d, as claimed.

In fact, we will see in the following section that the integer d is the negative of the arithmetic genus of X, the arithmetic genus being a natural extension to schemes of dimension 1 over a field of the classical notion of the genus of a curve. As it turns out, every projective double line of genus $-d$, with $d \geq 0$, is isomorphic to X.

There are also double lines of positive arithmetic genus—the simplest example of which is the "double conic" $\operatorname{Proj} k[x, y, z]/(xy - z^2)^2$, which has genus 3—and even continuous families of these when the genus is greater than or equal to 7. These objects arise naturally in the study of smooth curves: as a smooth nonhyperelliptic curve degenerates to a hyperelliptic curve, a phenomenon well known in the classical theory of varieties, the canonical model of the smooth curve approaches a projective double line (see Bayer and Eisenbud [1991] and Fong [1991] for more details).

EXERCISE III-27 ____

What is the ring structure of $\mathcal{O}_X(X)$ for the double line X above?

EXERCISE III-28* ____

Compute $\mathcal{O}_X(X)$ for the double line

$$X = \operatorname{Proj} k[u, v, x, y]/(x^2, xy, y^2, p(u, v)x + q(u, v)y)$$

where p and q are any homogeneous polynomials of degree d without common zeros other than $(0, 0)$. Prove that this double line is isomorphic to the double line of the example (and thus does not depend on the choice of p and q).

C   Invariants of Projective Schemes

In this section we assume that k is a field and work with k-schemes, except when explicit mention is made to the contrary.

Suppose that we are given a scheme in a projective space; how can we find invariants of it? The simplest idea is to ask: how many independent forms of degree d vanish on it? Putting the answers together, for various d, we get what used to be called the "postulation" of the scheme (presumably because one was then interested in schemes for which one postulated certain values for these numbers). Nowadays, it is usual to discuss this information in the equivalent form of the "Hilbert function." We will discuss here several variations of the method of Hilbert functions, which yield a wide range of invariants. Some of the invariants that we produce actually depend only on the abstract scheme and not on the given projective embedding, while others depend on the data associated to the embedding; and we will comment on these matters along the way. The approach we follow is the original one used by Hilbert (1890), rather than that of Samuel, which is more commonly adopted (see, for example, Hartshorne [1977, Chap. I]). Hilbert's method requires slightly more technique but yields a stronger and more easily understood result.

We begin by defining the basic invariants. In the last part of the chapter we will exhibit a number of simple geometric examples showing what sort of information the invariants contain.

i Hilbert Functions and Hilbert Polynomials

To begin with, suppose that we are given a closed subscheme $X \subset \mathbf{P}^r_k$, described by a saturated ideal $I = I(X) \subset S = k[x_0, \ldots, x_r]$ defined as in Example III-21. Suppose that the homogeneous polynomials $F_1, \ldots, F_n$ generate I. Write $R = S/I(X)$ for the homogeneous coordinate ring of X, and write R_v for the component of degree v.

The basic idea is to associate to $X \subset \mathbf{P}^r_k$ a function

$$H(X, \cdot) : \mathbf{N} \to \mathbf{N}$$

called the *Hilbert function of* X and defined by

$$H(X, v) = \dim_k R_v$$

More generally, if M is any finitely generated graded S-module, we define its Hilbert function to be $H(M, v) := \dim_k M_v$. The fundamental result is as follows.

THEOREM III-29   (Hilbert.) There exists a unique polynomial $P(X, v)$ in v such that $H(X, v) = P(X, v)$ for all sufficiently large v. More generally, for any finitely generated graded S-module M there exists a unique polynomial $P(M, v)$ such that $H(M, v) = P(M, v)$ for all sufficiently large v.

We will indicate below how this may be proved (along the lines of Hilbert's original proof [1890]).

The polynomial $P(X, v)$ is called the *Hilbert polynomial* of X. As in the classical case of varieties, it carries basic information about the scheme X. For example, we will see that its degree is the dimension of X, and in case X is of dimension 0, its (constant) value is the degree of X. More generally, we can define the degree of an n-dimensional subscheme X of projective space over a field k to be $n!$ times the leading coefficient of the Hilbert polynomial of X; this allows us to extend to the larger class of subschemes $X \subset \mathbf{P}^r$ the classical notion of degree for varieties. We will discuss this and other data conveyed by $P(X, v)$ at the end of Section III-C-iii and in Section III-C-iv.

ii Flatness III: Families of Projective Schemes

Another aspect of the significance of the Hilbert polynomial is that it gives us a geometric interpretation of the notion of flatness.

PROPOSITION III-30   A family of closed subschemes of a projective space over a reduced base is flat, in the sense introduced in Chapter II, if all fibers have the same Hilbert polynomial.

A proof of this in the general case would take us too far afield, but the result is easy when the base is $B = \operatorname{Spec} k[t]_{(t)}$.

Proof in case $B = \operatorname{Spec} k[t]_{(t)}$: A closed subscheme $X \subset \mathbf{P}^r_k \times B$ is given by an ideal I in

$$k[t]_{(t)}[x_0, \ldots, x_r]$$

which is homogeneous in $x_0, \ldots, x_r$. Thus each graded piece of the homogeneous coordinate ring

$$R = k[t]_{(t)}[x_0, \ldots, x_r]/I$$

is a module over $k[t]_{(t)}$.

As we know, the family $X \to B$ is flat iff each local ring $\mathcal{O}_{X,x}$ is $k[t]_{(t)}$-torsion-free. This is the same as saying that the torsion submodule of R goes to zero if we invert any of the x_i. It follows that the torsion submodule is killed by a power of the ideal $(x_0, \ldots, x_r)$ and thus meets only finitely many graded components of R. But if R_v is a graded component of R, then since $k[t]_{(t)}$ is a principal ideal ring and R_v is finitely generated as a $k[t]_{(t)}$-module, R_v is torsion-free iff it is free. Further, R_v is free if the number of generators it requires, which by Nakayama's lemma is

$$\dim_k R_v \otimes_{k[t]_{(t)}} k$$

is equal to its rank

$$\dim_{k(t)} R_v \otimes_{k[t]_{(t)}} k(t)$$

that is, iff the value of the $H(X_{(0)}, v)$ is equal to the value of $H(X_{(t)}, v)$, where $X_{(0)}$ and $X_{(t)}$ are the fibers of the family X over the two points (0) and (t) of B. (By the same argument, the Hilbert function itself is constant iff the family of affine cones $\mathrm{Spec}\, R$ is a flat family over B.) $\square$

This proposition shows that flat limits of closed subschemes of projective space behave better than flat limits in general. For example, though we have seen in Exercise II-24 that the flat limit of nonempty subschemes of an affine scheme may be empty, the proposition shows that this is not possible for flat limits of nonempty subschemes of a projective space. This, together with the existence and uniqueness of flat limits of closed subschemes in a one-parameter family (Exercise II-27), gives one approach to proving that projective schemes are proper, using the "valuative criterion." For all this, see, for example, Hartshorne (1977, Chap. II).

Of course $H(X, v)$ contains more information than $P(X, v)$, but it may appear that $P(X, v)$, as a polynomial with only finitely many coefficients, is easier to manipulate than the whole Hilbert function. Actually, the Hilbert function has a finite expression too, in terms of binomial coefficients. To see this, we will introduce a still finer set of invariants, the *graded betti numbers of the free resolution* of R, in terms of which both the Hilbert function and the Hilbert polynomial can be written conveniently. (The real advantage that the Hilbert polynomial has over the Hilbert function is that the information it contains depends a little less—in a sense we will make precise—on the details of the embedding of X.)

iii Free Resolutions

We will write $S(-b)$ for the graded, free module of rank 1 with generator in degree b; the apparently unfortunate choice of sign is recompensed by the convenient and eminently memorable formula

$$S(-b)_v = S_{v-b}$$

We can resolve R, or indeed any graded S-module, by using graded, free modules, which are direct sums of copies of modules of the form $S(-b)$. Here is how.

Let us suppose that $F_1, \ldots, F_n$ is a minimal set of generators for M. We will write $b_{0,j}$ for the degree of F_j. We define an epimorphism

$$\varphi_0 : E_0 := \bigoplus_{j=1}^{n} S(-b_{0,j}) \to M$$

by sending the generator of $S(-b_{0,j})$ to $F_j \in M$. Let $M^{(1)}$ be the kernel of φ_0. If $M^{(1)} \neq 0$, we repeat the process above with $M^{(1)}$ in place of M (which could be called $M^{(0)}$); choosing a minimal set of homogeneous elements $e_i^{(1)}$ of E_0 that generate $M^{(1)}$, with degrees $b_{1,i}$, we map a graded free module with generators of degrees $b_{1,i}$ onto M_1, by a map

$$\varphi_1 : E_1 := \bigoplus_{j=1}^{m} S(-b_{1,j}) \to E_0$$

sending the ith generator of E_1 to $e_i^{(1)}$. Continuing in this way, we obtain a resolution

$$\mathbf{E} : \cdots \to E_i \overset{\varphi_i}{\to} E_{i-1} \to \cdots \overset{\varphi_1}{\to} E_0$$

with

$$E_i = \bigoplus_j S(-b_{ij})$$

Of course, the process stops if some φ_i is a monomorphism. Hilbert's fundamental discovery was that this always occurs if S is a polynomial ring.

THEOREM III-31 (Hilbert's syzygy theorem.)   Let $S = k[x_0, \ldots, x_r]$. In any minimal free resolution as above, φ_i is a monomorphism for some $i \leq r + 1$, the number of variables; in particular, any graded S-module has a *finite*, graded, free resolution.

We will not prove this here; see Hilbert (1890) or, for a modern account, Matsumura (1986, Theorem 19.5).

The syzygy theorem allows us to prove Theorem III-29.

PROOF OF THEOREM III-29   The Hilbert function of an $S(-b)$ is easy to write down. Since

$$S(-b)_v = S_{v-b}$$

has a basis consisting of all monomials of degree $v - b$ in $r + 1$ variables, we see that

$$H(S(-b), v) = \binom{r + v - b}{r}$$

where the binomial coefficient is to be interpreted as 0 when the bottom is larger than the top. For $v \geq b - r$ this agrees with the polynomial

$$P(S(-b), v) = \frac{(r + v - b)(r + v - b - 1) \cdot \ \cdots \ \cdot (v - b)}{r(r - 1) \cdot \ \cdots \ \cdot 1}$$

so we see that $H(X, v)$ is a polynomial, for large v.

From a finite, free resolution for M as an S-module

$$E : 0 \longrightarrow E_{r+1} \overset{\varphi_{r+1}}{\longrightarrow} E_r \longrightarrow \cdots \longrightarrow E_1 \longrightarrow M \longrightarrow 0$$

with

$$E_i = \bigoplus_j S(-b_{ij})$$

we see that the Hilbert function of M can be written in the form

$$H(M, v) = \sum_{i=0}^{r} (-1)^i H(E_i, v) = \sum_{i=0}^{r} (-1)^i \sum_j H(S(-b_{ij}), v)$$

Since we have already shown that each $H(S(-b_{ij}), v)$ is a polynomial for large v, we see that $H(M, v)$ is a polynomial, for large v, as required. This proves Theorem III-29. $\qquad\square$

The Hilbert function and polynomial are clearly invariants of $X \subset \mathbf{P}_k^r$, but it is perhaps not obvious that the *graded betti numbers* b_{ij} are too. This follows from Nakayama's lemma; see, for example, Matsumura (1986, Section 19) for a discussion of minimal free resolutions over a local ring that translates immediately to the graded case.

We have thus three progressively weaker sets of invariants of a projective scheme: the graded betti numbers, the Hilbert function, and the Hilbert polynomial. To orient the reader, we will list some facts about them that we will not prove here and that will not be used in an essential way. Then we will give some examples.

1. As we have already mentioned, the degree d of the polynomial $P(X, v)$ is the dimension of X. The leading term is of the form

$$\frac{\delta(X)}{d!} v^d$$

and $\delta(X)$ is called the *degree* of X. It may be identified with the length of the subscheme in which X meets a general plane in $\mathbf{P}^r$ of dimension $r - d$. (See, for example, Hartshorne [1977, Chap. I, 7.5 and 7.7].) This follows from the observation, proved below (Proposition III-33), that the Hilbert polynomial of a zero-dimensional subscheme of length δ in $\mathbf{P}^n$ is the constant polynomial

δ together with the fact that if Y is a general hyperplane section of X, then the Hilbert polynomial of Y is the first difference function of the Hilbert polynomial of X—that is,

$$P(Y, v) = P(X, v) - P(X, v - 1)$$

2. (For readers who know about the relation between projective embeddings of X and invertible sheaves and also about cohomology; see, for example, Hartshorne [1977, Chap. III, exercises 5.2 and 5.3].) $P(X, v)$ depends only on X and the restriction $\mathscr{L}$ to X of the invertible sheaf $\mathcal{O}_{\mathbf{P}^r}(1)$, not on the choice of sections generating the line bundle that is the restriction of $H^0(\mathcal{O}_{\mathbf{P}^r}(1))$ to X. In fact, $P(X, v)$ is equal, for all v, to the alternating sum of dimensions of cohomology groups

$$\Sigma(-1)^i \dim_k H^i(\mathscr{L}^{\otimes v})$$

In particular, $P(X, 0) = \Sigma(-1)^i \dim_k H^i(\mathcal{O}_X)$ is a number depending on X and not on the embedding! In case X is a nonsingular curve over the complex numbers—that is, a Riemann surface—the number

$$\dim_k H^1(\mathcal{O}_X) = g = 1 - P(X, 0)$$

is the *genus* of X, and $1 - P(X, 0)$ turns out to be the right notion of genus for any one-dimensional scheme. It is called the *arithmetic genus* of the scheme. In the case where the dimension d of X is greater than one, it was at first felt that the normal case was the case where $H^i(\mathcal{O}_X) = 0$ for $1 < i < d$ (and this cohomology group always vanishes for $i > d$), so the arithmetic genus of X was by analogy defined as $1 + (-1)^d P(X, 0)$.

3. The set of all varieties in $\mathbf{P}^r$ with Hilbert polynomial equal to a given polynomial turns out to be itself naturally a projective scheme, called the Hilbert scheme associated with the given polynomial. For example, the Hilbert scheme of all projective subschemes with the same Hilbert polynomial as a k-plane turns out to be the variety of all k-planes in $\mathbf{P}^r$, which may be identified with the Grassmannian variety of all $k + 1$-dimensional subspaces of a given $r + 1$-dimensional space. There are, however, not many other cases in which these Hilbert schemes have been understood geometrically! We will return to this construction in Sections IV-B-iv and IV-B-v of the next chapter.

EXERCISE III-32

Show that if A is a Noetherian ring and $\mathscr{X}$ is a closed subscheme of $\mathbf{P}^n_A$, regarded as a family of schemes over Spec A, then, writing X_p for the fiber over a point $p \in \operatorname{Spec} A$, the function $H(X_p, v)$, regarded as a function in p, is upper semicontinuous in the Zariski topology on $\operatorname{Spec} A$; that is, for any v and any number m,

$$\{ p \in \operatorname{Spec} A \,|\, H(X_p, v) \geq m \}$$

is a closed subset of $\operatorname{Spec} A$.

4. In many ways the invariant provided by the graded betti numbers is the most subtle of all, and until very recently nothing was known of its geometric significance beyond that of the Hilbert function and polynomial. Now, however, we know in a few cases (and conjecture in a few more) how they reflect some subtle aspects of the intrinsic geometry of X. See, for example, the work of Green (1984) and Green and Lazarsfeld (1985) for more information.

iv Examples

a Points in the Plane

Already for the case of zero-dimensional subschemes in the plane we get different information from the Hilbert polynomial, Hilbert function, and graded betti numbers.

First of all, we have stated above that the Hilbert polynomial of a subscheme $X \subset \mathbf{P}^r$ is a polynomial whose degree is equal to the dimension of X; so when X is zero-dimensional, the Hilbert polynomial is a constant. We can easily prove this and somewhat more in the case of points.

PROPOSITION III-33   The Hilbert function of a zero-dimensional subscheme of length δ in $\mathbf{P}^r$ satisfies $H(X, v) \leq \delta$ for all v, with equality for large v. Thus $P(X, v) \equiv \delta$.

Proof   We must show that the codimension in $k[x_0, \ldots, x_r]$ of the set of homogeneous forms of degree v that vanish on X—that is, $\operatorname{codim} I(X)_v$—is less than or equal to δ, with equality for large v. The reason is that vanishing at a point is one linear condition on the coefficients of a polynomial, and thus vanishing at X should be δ linear conditions; for large v we will show that these conditions are always linearly independent.

To make this precise, we pass to an affine open set. Changing coordinates, we may suppose that X is contained in the affine open set $x_r \neq 0$, so that a form F of degree v belongs to $I(X)$ iff $F(x_0, \ldots, x_{r-1}, 1)$ belongs to the ideal $J \subset k[x_0, \ldots, x_{r-1}]$ of X in the affine open set $x_r \neq 0$. To say that X is of length δ means that J is of codimension δ in $k[x_0, \ldots, x_{r-1}]$ and thus of codimension less than or equal to δ in the space of those polynomials that can be written as $F(x_0, \ldots, x_{r-1}, 1)$ for F of degree v—these are simply the polynomials in $k[x_0, \ldots, x_{r-1}]$ of degree less than or equal to v. This shows at once that $H(X, v) \leq \delta$ for all v, with equality if J has codimension δ in the space of polynomials of degree less than or equal to v. But J will have codimension δ in the space of polynomials of degree less than or equal to v as soon as a set of representatives for $k[x_0, \ldots, x_{r-1}]/J$ can be chosen from among the polynomials of degree less than or equal to v, which is certainly true for all large v.

$\square$

We remark that if $X \subset \mathbf{P}^r$ is nonempty, then $I(X)$ contains nothing of degree 0 (remember that we are working over a field!), so $H(X, 0) = 1$. Thus the proposition provides easy examples where $P(X, 0) \neq H(X, 0)$.

We can easily exhibit a family of subschemes of $\mathbf{P}^2_k$ with constant Hilbert polynomial but varying Hilbert function. To construct such a family $\mathscr{X} \subset \mathbf{P}^2_k \times \operatorname{Spec} k[t]$, for example, we can take the "constant" points P and Q given by $(x_2 = x_1 + x_0 = 0)$ and $(x_2 = x_1 - x_0 = 0)$, and the variable point R given by $(x_1 = x_2 - t \cdot x_0 = 0)$, and let $\mathscr{X}$ be the (disjoint) union of P, Q, and R in $\mathbf{P}^2_k \times \operatorname{Spec} k[t]$.

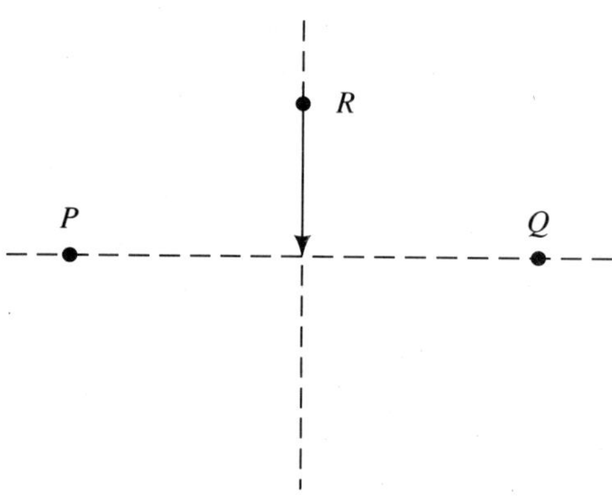

We regard $\mathscr{X}$ as a flat family over $\operatorname{Spec} k[t]$ by means of the projection to the second factor, whose fibers $X_{(0)}$ and X_λ over the generic point and over every closed point $(t - \lambda)$ (as schemes over $k(t)$ and k, respectively) have Hilbert polynomials $P(X_{(0)}, v) = P(X_\lambda, v) \equiv 3$; but while the Hilbert function $H(X_\lambda, 1) = 3$ for $\lambda \neq 0$, we have $H(X_0, 1) = 2$.

EXERCISE III-34

Let $\widetilde{X}_\lambda \subset \mathbf{A}^3$ be the cone over the fiber X_λ of the family $\mathscr{X}$ above. Show that there does not exist a flat family $\widetilde{\mathscr{X}} \subset \mathbf{A}^3 \times \operatorname{Spec} k[t]$ whose fiber over each point $(t - \lambda)$ is $\widetilde{X}_\lambda$. (There does exist such a family over the complement of the origin in $\operatorname{Spec} k[t]$, however.) What is the flat limit of the cones $\tilde{X}_\lambda$ as λ approaches 0? (See Exercise III-41.)

Now consider the case where X is a set of four distinct points in the plane $\mathbf{P}_k^2$. We already know that $P(X,v) \equiv 4$. We will treat separately the cases where all the points—or all but one of the points—lie on a line.

1. *The case where X is contained in a line.*

Suppose, first, that the points lie on a line L, with equation $l = 0$, say. The only line containing X is L, so $H(X,1) = H(\mathbf{P}_k^2,1) - 1 = 2$. If $q = 0$ is the equation of a conic containing X, then q restricts to a form of degree 2 on L, vanishing at the four points of X, so q must vanish identically on L. Thus $q = 0$ is the union of L and one other line, and the set of equations of conics containing X is the three-dimensional space of multiples of l by linear forms. This gives $H(X,2) = H(\mathbf{P}_k^2,2) - 3 = 3$.

Starting with $v = 3$, however, vanishing at the four points imposes four independent conditions on forms of degree v, so $H(X,v) = 4$. To prove this, it is enough, for each 3 point subset X' of X, to find a curve of degree v that contains X' but not the fourth point of X. We may do this with a curve consisting of v straight lines, three of these passing through one each of the points of X' and the rest far away from X:

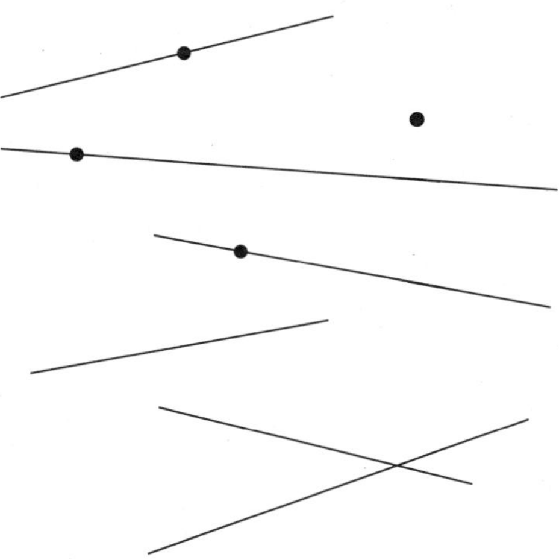

To compute the minimal free resolution in this and the next examples, we will use a result of Hilbert, which was generalized and extended to the local case by Lindsay Burch.

THEOREM III-35 If I is the homogeneous ideal of a zero-dimensional subscheme $X \subset \mathbf{P}_k^2$, then any minimal free resolution of the homogeneous coordinate ring S/I has the form

$$0 \to \sum_{j=1}^{n-1} S(-b_{2j}) \xrightarrow{A} \sum_{j=1}^{n} S(-b_{1j}) \to S$$

Further, the jth generator of I—that is, the image of $S(-b_{1j})$ in S—is up to sign the determinant of the matrix A with the jth row deleted.

For a proof, see, for example, Eisenbud (1991).

We will make use of this to compute minimal generators of the ideal $I(X)$ through the following corollary.

COROLLARY III-36   If I is the homogeneous ideal of a zero-dimensional subscheme $X \subset \mathbf{P}_k^2$, and if I contains an element of degree e, then I can be generated by $e + 1$ elements.

Proof   If the minimal number of generators of I is g, then I is generated by $(g - 1) \times (g - 1)$ determinants of a matrix A whose entries are in the graded maximal ideal of S and are thus forms of positive degree. Consequently, no element of I has degree less than $g - 1$, and we have $g \le e + 1$, as claimed.

$\square$

Note that by the theorem, knowing the degrees of the entries of the matrix A is equivalent to knowing the graded betti numbers in this case: for the b_{1j} are just the degrees of the minors of A, and b_{2j} is the sum of b_{1i} plus the degree of the ijth entry of A.

Applying this to the example at hand, we see that since X lies on a line, $I(X)$ may be generated by two elements, which may, of course, be taken to be L and a form of smallest possible degree in I that is not divisible by L.

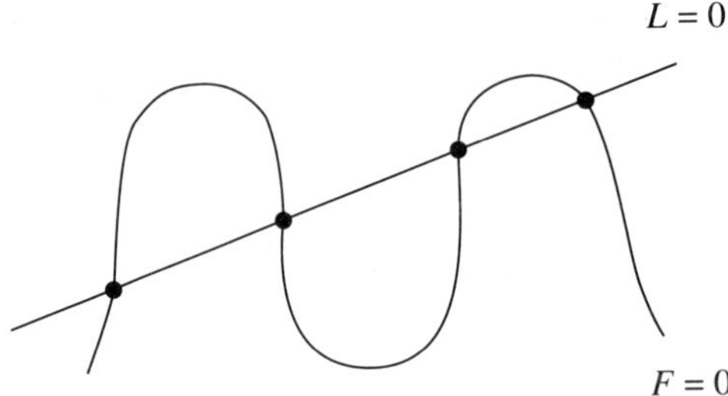

As we have noted, this smallest possible degree is 4, and we may, for example, take the quartic form F, where $F = 0$ is the equation of the quartic consisting of four lines, each through one point of X.

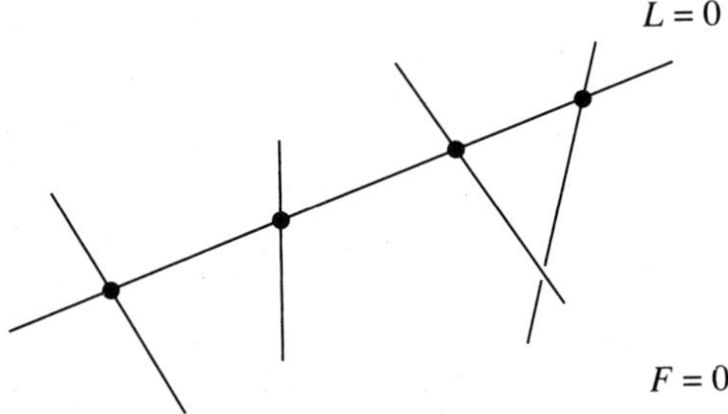

Since L and F have no common factor, we see that the minimal free resolution of $S/I(X)$ has the form

$$0 \to S(-5) \to S(-4) \oplus S(-1) \to S$$

giving the expression for the Hilbert function

$$H(X, v) = \binom{v+2}{2} - \binom{v+1}{2} - \binom{v-2}{2} + \binom{v-3}{2}$$

2. The case where all but one of the points of X lie on a line.

Next, consider the case where only three of the four points lie on the line L. Now there is no linear form in $I(X)$, so $H(X, 1) = 3$.

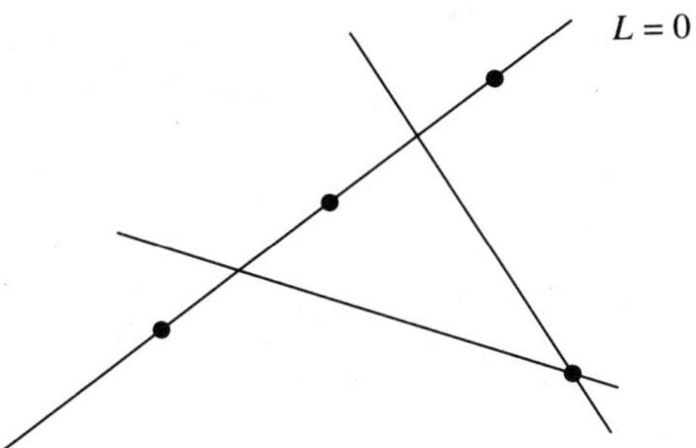

Any quadric containing the three points on L must, by the same argument as before, contain L; so any quadric containing X is the union of L and a line through the fourth point. Since the space of linear forms corresponding to lines through the fourth point is two-dimensional, the space of quadrics containing X is two-dimensional; and we have $H(X, 2) = 4$. Following the same argument as before, we show that $H(X, v) = 4$ for all larger v, so this is the case for all $v \geq 2$.

As for the resolution, we see by the corollary above that $I(X)$ requires at most three generators. But $I(X)$ is *not* generated by the two independent quadrics it contains, since these have a common factor; thus it is minimally generated by these two quadrics and another generator, an element of smallest possible degree not contained in the ideal generated by the two quadrics or, equivalently, vanishing on a curve not containing L. It is easy to see that there is a cubic curve with the desired properties; it may be taken, for example, to be the union of three lines, each passing through one of the points of L and one passing, in addition, through the fourth point. Since the minimal generators of $I(X)$ have degrees 2, 2, 3, the matrix A must be a 2×3 matrix whose entries have degrees as given in the following diagram (up to a rearrangement of the rows and columns):

$$A = \begin{pmatrix} 1 & 2 \\ 1 & 2 \\ 0 & 1 \end{pmatrix}$$

(where, of course, the entry of degree 0 must actually be 0, since all the entries must be in the maximal graded ideal). Thus the minimal free resolution has the form

$$0 \to S(-3) \oplus S(-4) \xrightarrow{A} S(-2) \oplus S(-2) \oplus S(-3) \to S$$

3. *The case where no three of the points of X lie on a line.*

Finally, consider the case where X consists of four points, no three of which lie on a line. We claim that the Hilbert function of X is the same as in the previous case: $H(X, 1) = 3$, $H(X, v) = 4$ for $v \geq 2$. The first of these values is obvious, since X lies on no lines. For the second, it is enough as before to note that there are quadrics (and thus *a fortiori* forms of higher degree) containing any subset of the four points but missing the last; these may be constructed as before as unions of lines.

Now we compute the free resolution of $S/I(X)$. Taking the two pairs of opposite sides of the quadrilateral formed by the points gives us two quadrics q_1 and q_2 without common factor in the ideal of X.

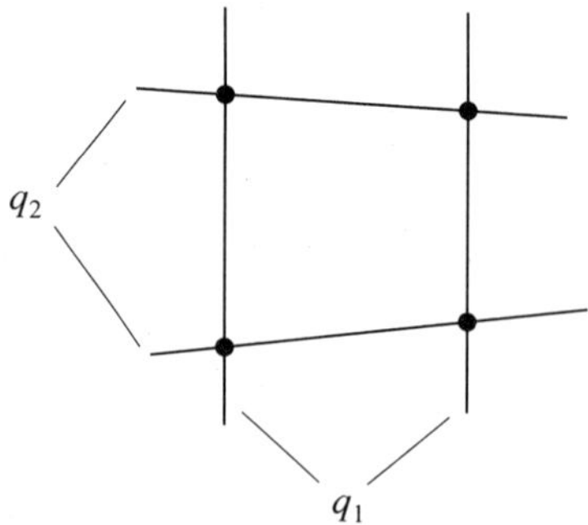

Since q_1 and q_2 are relatively prime, the free resolution of the ideal I *they* generate has the form

$$0 \to S(-4) \xrightarrow{A} S(-2) \oplus S(-2) \xrightarrow{B} S$$

where

$$B = (q_1 \quad q_2) \qquad A = \begin{pmatrix} -q_2 \\ q_1 \end{pmatrix}$$

Computing the Hilbert function of S/I from this resolution, we see that it is the same as that of $S/I(X)$, and since $I \subset I(X)$, we must have $I = I(X)$; that is, $I(X)$ is generated by q_1 and q_2 and X is correspondingly the intersection of the two conics containing it.

Summing up, we see that all three of the examples look the same from the point of view of Hilbert polynomials; the first two examples are distinguished by their Hilbert functions; and the last two examples look the same from the point of view of Hilbert functions but are distinguished by their graded betti numbers. It is not hard to find corresponding examples of subschemes X of length 4 where the properties distinguished are actually intrinsic properties of the schemes, not dependent on the embedding. For example, while the scheme $\operatorname{Spec} k[x]/(x^4)$ may be embedded in $\mathbb{P}^2$ so as to have any of the Hilbert functions and betti numbers above (for instance as the subschemes defined by the ideals (x_0, x_1^4), $(x_0 x_2^2 - x_1^3, x_0 x_1, x_0^2)$, and $(x_0 x_2 - x_1^2, x_0^2)$ respectively), the subscheme defined by (x_0^2, x_1^2) will always have the graded betti numbers and Hilbert function of case 3.

EXERCISE III-37

Find the Hilbert polynomial, Hilbert function, and graded betti numbers of all subschemes of the plane of length 3.

b Double lines in $\mathbf{P}_k^3$

Let X_n be the double line in $\mathbf{P}_k^3$ defined by the ideal

$$I_n = (x_0^2, x_0 x_1, x_1^2, x_0 x_2^n + x_1 x_3^n)$$

For each n the ideal I_n contains the ideal

$$I = (x_0^2, x_0 x_1, x_1^2)$$

Since S/I is a free $k[x_2, x_3]$-module on the generators 1, x_0, and x_1, we see that

$$H(S/I, v) = H(S/(x_0, x_1), v) + 2H(S/(x_0, x_1), v - 1)$$

Further, we see easily, using this basis, that if we write $p = x_0 x_2^n + x_1 x_3^n$ for the last generator of I_n, then for any homogeneous form $q = q(x_0, \ldots, x_3)$ we have $qp \in I$ iff $q \in (x_0, x_1)$. Thus

$$H(S/I_n, v) = H(S/I, v) - H(S/(x_0, x_1), v - n - 1)$$

But $P(S/(x_0, x_1), v) = v + 1$. Putting all these equalities together, we get

$$P(X_n, v) = 2v + n + 1$$

so the Hilbert polynomial, and even the arithmetic genus

$$p_a(X_n) = 1 - P(X_n, 0) = -n$$

distinguishes between these double lines.

Here is an exercise that will be useful for the following three examples.

EXERCISE III-38 ____

Compute the Hilbert polynomials of the following subschemes of $\mathbf{P}_k^3$:

a. The union of two skew lines.
b. The union of two incident lines.
c. The subscheme supported on the union of two incident lines with an embedded point of degree 1 at their point of intersection, not lying in the plane spanned by the two lines. Also, show that for any point p on the double line X_0 (or on any of the double lines above) there is a unique subscheme of $\mathbf{P}_k^3$ consisting of X_0 with an embedded point of degree 1 at p; and compute the Hilbert polynomial of this subscheme.

Given this exercise, we can use the notion of Hilbert polynomial to further illuminate the example of a family of pairs of skew lines tending to a pair of incident lines. Recall that in Section II-C-iii-b of Chapter II we discussed such a family and showed that the flat limit was not reduced: it was supported on the union of the incident lines but had an embedded point at their point of intersection. As the exercise above suggests, if we complete these families in $\mathbf{P}_k^3$, we see that this is necessary from the point of view of Hilbert polynomials.

Consider next a family of pairs of skew lines in $\mathbf{P}_k^3$, described as follows. First, let $L \subset \mathbf{P}_k^3$ be the constant line $(x_0 = x_1 = 0)$, and let $M_t \subset \mathbf{P}_k^3$ be the line $(x_0 = t \cdot x_3, x_1 = t \cdot x_2)$. Let Y_t be the union of these two lines. We may ask then for the flat limit of the family Y_t—in other words, the fiber Y_0 over the origin in $\mathbf{A}^1$ of the union $\mathscr{Y}$ of the subschemes $\mathscr{L}$ and $\mathscr{M}$ of $\mathbf{P}_k^3 \times \mathbf{A}^1$ given by $(x_0 = x_1 = 0)$ and $(x_0 = t \cdot x_3, x_1 = t \cdot x_2)$, respectively. Of course, the support of Y_0 will be the line L, but it is equally clear that it must have some nonreduced structure. In fact, the flat limit is none other than the double line X_1 above.

EXERCISE III-39

Verify that the flat limit Y_0 is the double line X_1. *Note:* By comparing Hilbert polynomials, it is enough to prove inclusion in one direction.

An interesting wrinkle on this last construction is to consider a slightly different family of pairs of skew lines: we let L be as above, and let M_t be the line given by $(x_0 = t \cdot x_3, x_1 = -t^2 \cdot x_2)$. At first glance it might appear that the flat limit of the unions $Y_t = L \cup M_t$ will be the double line given by $(x_0^2 = x_1 = 0)$, which is isomorphic to the double line X_0 above; but this cannot be, since the Hilbert polynomials are not equal. The following exercise gives the real situation.

EXERCISE III-40

Show that with L and M_t as above, the flat limit as $t \to 0$ of the union $L \cup M_t$ is the double line $(x_0^2 = x_1 = 0)$ with an embedded point of degree 1 located at the point $[0, 0, 1, 0]$.

EXERCISE III-41

Let L, M, and $N_t \subset \mathbf{P}_k^3$ be the lines $x_2 = x_3 = 0$, $x_1 = x_3 = 0$, and $x_1 + x_2 = t \cdot x_1 + (1 - t) \cdot x_3 = 0$, respectively; let Z_t be their union in $\mathbf{P}_k^3$. Find the Hilbert polynomial of Z_1 and the Hilbert polynomial of Z_0. What is the "limit," in the sense discussed above, of the subschemes $Z_t \subset \mathbf{P}_k^3$ as $t \to 0$?

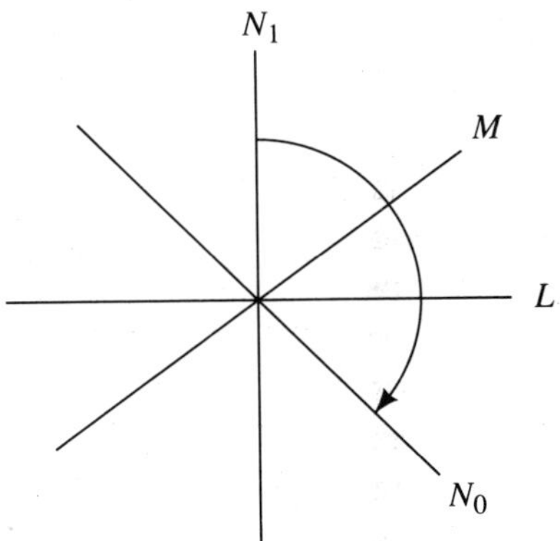

Finally, there are arithmetic analogues of each of the last three examples. For example, consider the following three families of subschemes of $\mathbf{P}^3_{\mathbf{Z}}$. (These are flat families over Spec $\mathbf{Z}$):

i. Let $\mathcal{L} \subset \mathbf{P}^3_{\mathbf{Z}}$ be the constant line ($x_0 = x_1 = 0$), and let $\mathcal{M} \subset \mathbf{P}^3_{\mathbf{Z}}$ be the line ($x_0 = 7 \cdot x_3, x_1 = -7 \cdot x_2$); let $\mathcal{Y}$ be the union of these two subschemes.

ii. Let $\mathcal{L} \subset \mathbf{P}^3_{\mathbf{Z}}$ be the constant line ($x_0 = x_1 = 0$), and let $\mathcal{M} \subset \mathbf{P}^3_{\mathbf{Z}}$ be the line ($x_0 = 7 \cdot x_3, x_1 = -49 \cdot x_2$); let $\mathcal{Y}$ be the union of these two subschemes.

iii. Let $\mathcal{L}$, $\mathcal{M}$, and $\mathcal{N} \subset \mathbf{P}^3_{\mathbf{Z}}$ be the subschemes defined by ($x_2 = x_3 = 0$), ($x_1 = x_3 = 0$), and ($x_1 + x_2 = 7 \cdot x_1 - 6 \cdot x_3 = 0$), respectively; let $\mathcal{Y}$ be their union.

EXERCISE III-42

For each of the subschemes $\mathcal{Y} \subset \mathbf{P}^3_{\mathbf{Z}}$ above, find the fiber of $\mathcal{Y}$ over the point $(7) \in \mathrm{Spec}\ \mathbf{Z}$. Compare your answer with that found in the preceding three exercises.

IV

The Functor of Points

In this chapter we will discuss a way of embedding the category of schemes into a larger category of functors. This apparent step toward vapid abstraction turns out to make many geometric arguments easy and feasible. We will discuss, by way of illustration, the applications of the idea to the geometry of schemes in the notions of

> open subfunctors,
> the arithmetic notion of the rational points of a variety,
> the tangent spaces to a variety,
> group schemes,

and to the construction, or really the specification, of a scheme, taking as our examples

> fibered products,
> projective space,
> the Hilbert scheme (and other moduli schemes).

We should mention briefly here two ways in which this chapter differs from those preceding it. The first is that the concepts introduced here are much less ubiquitous than those that have come before. There is hardly a paper in algebraic geometry today that does not use the language we have described in the preceding three chapters. By contrast, only a minority of such papers actually use the technique of the functor of points. It arises primarily in dealing with moduli and parameter spaces and group schemes (where it is fundamental) and also to some extent in arithmetic algebraic geometry. Thus this chapter is much less of a prerequisite for further study in algebraic geometry than the first three were. (It is also the case that while there exist many references for the topics discussed in Chapters I–III, there are fewer references for the material in this chapter—in particular, these topics are for the most part not discussed in Hartshorne.)

Secondly, because the interesting applications of the functor of points lie in deeper areas of algebraic geometry, the background necessary for understanding the details of this chapter is somewhat greater than for the preceding chapters.

For both these reasons, some readers may prefer simply to browse through this chapter on first reading.

A   The Functor of Points and the Geometry of Schemes

i The Functor of Points

One of the intriguing things about schemes is precisely that they have so much structure that is not conveyed by their underlying sets, so that the familiar operations on sets such as taking direct products require vigilant scrutiny lest they turn out not to make sense.

Part of the problem is that the points of a scheme do not in general look anything like one another: we have nonclosed points as well as closed ones; and if we are working over a non–algebraically closed field, then even closed points may be distinguished by having different residue fields. Similarly, if we are working over $\mathbf{Z}$, different points may have residue fields of different characteristic; and if we extend the notion of point to "closed subscheme whose underlying topological space is a point," we have an even greater variety. And, of course, a morphism between schemes will not at all be determined by the associated map on underlying point sets.

There is, however, a way of looking at a scheme—via its *functor of points*—that reduces it in effect to a set. More precisely, we may think of it as an organized collection of sets, a functor on the category of schemes, on which the familiar operations on sets behave as usual. In this section we will examine this functorial description. A big payoff is that we will see the category of schemes embedded in a larger category of functors, in which many constructions are much easier. The advantage of this is something like the advantage in analysis of working with distributions, not just ordinary functions; it shifts the problem of making constructions in the category of schemes to the problem of understanding which functors come from schemes. Further, many geometric constructions that arise in the category of schemes can be extended to larger categories of functors in a useful way.

To introduce the notion of the functor of points, we start out in a general categorical setting. To begin with, in many categories whose objects are sets with additional structure, the underlying set $\#X$ of an object X may be described as the set of morphisms from a universal object to X; for example:

i. In the category of differentiable manifolds, if Z is the manifold consisting of one point, then for any manifold X we have $\# X = \text{Hom}(Z, X)$.

ii. In the category of groups, if Z is the group $\mathbf{Z}$ of integers, we have for any group X, $\# X = \text{Hom}(Z, X)$.

iii. In the category of rings with unit and unit-preserving homomorphisms, if we set $Z = \mathbf{Z}[x]$, then for any ring X with unit we have $\# X = \text{Hom}(Z, X)$.

In general, for any object Z of a category $\mathscr{X}$, the association

$$X \mapsto \text{Hom}_{\mathscr{X}}(Z, X)$$

defines a functor φ from the category $\mathscr{X}$ to the category of sets. As indicated in the first paragraph above, however, it is not really satisfactory to call the set $\varphi(X) = \text{Hom}_{\mathscr{X}}(Z, X)$ the set of points of the object X unless this functor is *faithful*—that is, unless for any pair of objects X_1 and X_2 of $\mathscr{X}$ a morphism

$$f : X_1 \to X_2$$

is determined by the map of sets

$$f' : \text{Hom}_{\mathscr{X}}(Z, X_1) \to \text{Hom}_{\mathscr{X}}(Z, X_2)$$

It may not always be possible to satisfy this condition. For example, let (Hot) be the category of CW-complexes, where $\text{Hom}_{(\text{Hot})}(X, Z)$ is the set of homotopy classes of continuous maps from X to Z. If Z is the one-point complex, then

$$\text{Hom}_{(\text{Hot})}(Z, X) = \pi_0(X)$$

the set of connected components of X, and this does not give a faithful functor. Nor is it possible to chose a better object Z. Likewise, in the category of schemes, there is no one object Z that will serve in this capacity.

Grothendieck's ingenious idea was to remedy this situation by considering not just one set $\text{Morph}(Z, X)$ but all at once! That is, we associate to each scheme X the "structured set" consisting of all the sets $\text{Morph}(Z, X)$, together with, for each morphism, $f : Z \to Z'$, the mapping from $\text{Morph}(Z', X)$ obtained by composing with f.

To put this more formally, the *functor of points* of a scheme X is the "representable" functor determined by X; that is, the functor

$$h_X : (\text{schemes})^o \to (\text{sets})$$

where (schemes)^o and (sets) represent the category of schemes with the arrows reversed and the category of sets, respectively; h_X takes each scheme Y to the set

$$h_X(Y) = \text{Morph}(Y, X)$$

and each morphism $f : Y \to Z$ to the map of sets

$$h_X(Z) \to h_X(Y)$$

defined by sending an element $g \in h_X(Z) = \text{Morph}(Z, X)$ to the composition $g \circ f \in \text{Morph}(Y, X)$. The reason for the name "representable functor" is that we say this functor is *represented* by the scheme X. The set $h_X(Y)$ is called the set of *Y-valued points of* X (if $Y = \text{Spec } T$ is affine, we will often write $h_X(T)$ instead of $h_X(\text{Spec } T)$ and call it the set of T-valued points of X).

To introduce one more layer of abstraction, note that this construction defines a functor

$$h : (\text{schemes}) \to \text{Funct}((\text{schemes})^o, (\text{sets}))$$

(where morphisms in the category of functors are natural transformations), sending

$$X \mapsto h_X$$

and associating to a morphism $f : X \to X'$ the natural transformation $h_{X'} \to h_X$ that for any scheme Y sends $g \in h_{X'}(Y) = \text{Morph}(X', Y)$ to the composition $f \circ g \in \text{Morph}(X, Y)$.

Of course, when we want to work with schemes over a given base S, we should take morphisms over S as well. The situation is completely analogous to that above: we describe in this way a functor

$$X \mapsto h_X$$

from the category of S-schemes to the category

$$\text{Funct}((S\text{-schemes})^o, (\text{sets}))$$

While in most of the sequel we will simply refer to the functor of points of a scheme X, it is important to remember that if we want to do geometry over a scheme S—for example, if we want to work over a particular field such as the complex numbers—we have to take the functor of points in this sense, replacing $\text{Morph}(Y, X)$ by $\text{Morph}_S(Y, X)$. We will see this below in an exercise.

The functor of points h_X really does determine the scheme X. This follows from a basic categorical fact.

LEMMA IV-1 (Yoneda's Lemma.)   Let $\mathscr{C}$ be a category and let X, X' be objects of $\mathscr{C}$.

   a. If F is any contravariant functor from $\mathscr{C}$ to the category of sets, then the maps of functors from $\mathrm{Hom}(-, X)$ to F are in natural correspondence with the elements of $F(X)$.

   b. If the functors $\mathrm{Hom}(-, X)$ and $\mathrm{Hom}(-, X')$ from $\mathscr{C}$ to the category of sets are isomorphic, then $X \cong X'$. More generally, the maps of functors from $\mathrm{Hom}(-, X)$ to $\mathrm{Hom}(-, X')$ are the same as maps from X to X'; that is, the functor

$$h : \mathscr{C} \to \mathrm{Funct}(\mathscr{C}^0, (\mathrm{sets}))$$

sending

$$X \mapsto h_X$$

is an equivalence of $\mathscr{C}$ with a full subcategory of the category of functors.

Proof   a. The correspondence sends $\alpha : \mathrm{Hom}(-, X) \to F$ to the element $\alpha(1_X)$, where $1_X : X \to X$ is the identity map. The inverse takes $p \in F(X)$ to the map α sending an element $f \in \mathrm{Hom}(Y, X)$ to $F(f)(p) \in F(Y)$.

   b. Apply the statement of part *a* to the functor $F = \mathrm{Hom}(-, Y)$. $\square$

The following improvement of Lemma IV-1 shows that it is enough to look at the functor of points restricted to the category of affine schemes, or, equivalently, to the category $(\mathrm{rings})^0$, the category of commutative rings with the arrows reversed; and the same thing works in the relative setting.

PROPOSITION IV-2   If S is a commutative ring, then a scheme over S is determined by the restriction of its functor of points to affine schemes over S; in fact

$$h : (S\text{-schemes}) \to \mathrm{Funct}((S\text{-algebras}), (\mathrm{sets}))$$

is an equivalence of the category of S-schemes with a full subcategory of the category of functors.

Of course, a contravariant functor on the category of affine schemes is the same as a covariant functor on the category of rings; so given this result, we will generally think of our contravariant representable functors h_X as covariant functors on S-algebras.

Proof   This is really just the statement that schemes are built up out of affine schemes. Write h_X for the functor $\mathrm{Morph}_S(\cdot, X)$ restricted to the category of affine schemes over S. It is enough to show that any natural transformation $\varphi : h_X \to h_{X'}$ comes from a unique morphism f over S from X to X'. To construct f from φ, let $\{U_a\}$ be an affine cover of X, and apply φ to the inclusion maps $U_a \subset X$ to get morphisms $U_a \to X'$. These will satisfy the compatibility conditions necessary to define the desired morphism f. Uniqueness comes down to the observation that two morphisms from X to X' that differ are already different when restricted to one of the U_a. $\square$

EXERCISE IV-3

Suppose that X is (like virtually all schemes of interest to us) *locally Noetherian*—that is, covered by spectra of Noetherian rings. Prove that X is determined by the restriction of h_X to the category of Noetherian local rings and local homomorphisms.

The apparently abstract idea of the functor of points has its root in the study of solutions of equations. Let $X = \mathrm{Spec}\, R$ be an affine scheme. We may present R as $R = \mathbf{Z}[x_1, x_2, \ldots]/(f_1, f_2, \ldots)$. If T is any other ring (one should think of $T = \mathbf{Z}, \mathbf{Z}/p, \mathbf{Z}_p, \hat{\mathbf{Z}}_p, \mathbf{Q}_p, \hat{\mathbf{Q}}_p, \mathbf{R}, \mathbf{C}, \ldots$), then a morphism from $\mathrm{Spec}\, T$ to $\mathrm{Spec}\, R$ is the same as a ring homomorphism from R to T, and this is determined by the images a_i of the x_i. Of course, a set of elements $a_i \in T$ determines a morphism in this way iff they are solutions to the equations $f_i = 0$. We have shown

$$h_X(T) = \{\text{sequences of elements } a_1, \ldots \in T \text{ that are solutions of}$$
$$\text{the equations } f_i = 0\}$$

Similarly, if X is an arbitrary scheme, so that X is the union of affine schemes X_a meeting along open subsets, then a map from an affine scheme Y to X may be described by giving a covering of Y by distinguished affine open subsets Y_{f_a} and maps from Y_{f_a} to X_a for each a, agreeing on open sets (some of the Y_{f_a} may, of course, be empty). Thus an element of $h_X(Y)$ may be described even in this general context as a set of solutions to systems of equations, corresponding to some of the X_a, with compatibility conditions satisfied by the solutions on the sets where certain polynomials are non-zero.

Even with this interpretation, the notion of the functor of points may seem at this point an arid one: while we can phrase problems in this new language, it's far from clear that we can solve them in it. The key to being able to work in this setting is the fact that many apparently geometric notions have natural extensions from the category of schemes to larger categories of functors. In the remainder of this section we will illustrate this with a few examples.

ii Open Subfunctors

In this example we show how to define an open subfunctor of a functor

$$F \in \mathrm{Funct}((\mathrm{rings}) \to (\mathrm{sets}))$$

The idea is that an open subscheme is a "sub-something" that when intersected with any affine open subscheme is an open subset of the affine scheme. Thus we begin by computing the functor that represents an open subscheme of an affine scheme, say $U \subset \mathrm{Spec}\, R$. Suppose that the complement of U is the closed set $V(I)$ with $I \subset R$. For any ring T, the morphisms of $\mathrm{Spec}\, T$ to U are the morphisms of $\mathrm{Spec}\, T$ to $\mathrm{Spec}\, R$ that factor through U; that is,

$$h_U(T) = \{\varphi \in h_{\mathrm{Spec}\, R}(T) \,|\, \varphi^*(I)T = T\}$$

In general, a map of functors $\alpha : G \to F$ is an open subfunctor iff the following hold:

1. α is injective; that is, for all rings T,

$$G(T) \to F(T)$$

   is an injection.
2. For every ring R and every element $\psi \in F(R)$, the fibered product

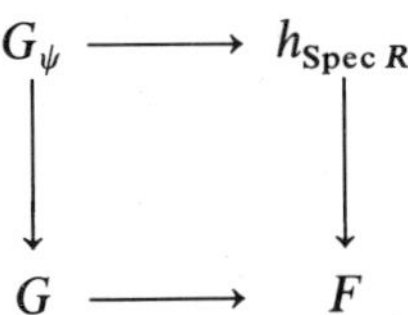

of functors yields a map $G_\psi \to h_{\mathrm{Spec}\, R}$ that is isomorphic to an open subscheme of $h_{\mathrm{Spec}\, R}$.

EXERCISE IV-4 ____

Let X be a scheme, and let Y be an open subscheme. Show that h_Y is indeed an open subfunctor of h_X.

EXERCISE IV-5 ____

Let X be a scheme over the field k. Define a functor $F : (\text{schemes}/k) \to (\text{sets})$ as follows: for each scheme Y, let $F(Y)$ be the set of closed subschemes $Z \subset X \times Y$, flat over Y, such that all the fibers of Z over closed points of Y are subschemes of degree 2 of X. Let G be the subfunctor of F obtained by adding the requirement that the fibers of Z over closed points of Y are reduced. Show that $G \subset F$ is open.

A closed functor is similarly a subfunctor that is the complement—object by object—of an open subfunctor. One warning: imposing what may appear to be closed conditions on a functor does not necessarily make a closed subfunctor, as the following exercise shows.

EXERCISE IV-6 ____

Let X be a scheme over the field k. Define the functor $F : (\text{schemes}/k) \to (\text{sets})$ as in Exercise IV-5, and define a subfunctor H of F by letting, for each scheme Y, $H(Y)$ be the set of closed subschemes $Z \subset X \times Y$, flat over Y, such that all the fibers of Z over closed points of Y are subschemes of degree 2 of X supported at a single point of X. Show that H is not in general a closed subfunctor of F.

iii *k*-Rational Points

If X is a scheme over a field k, then the k-valued points of X over k are maps $\operatorname{Spec} k \to X$ whose composition with the natural map $X \to \operatorname{Spec} k$ is the identity. We claim that such maps correspond exactly to closed points p of X that are *rational over k* (or "k-rational") in the sense that the residue class field $\kappa(p)$ is k (via the inclusion map of k into the local ring of X at p). Indeed, since $\operatorname{Spec} k$ has no nontrivial open coverings, a map from $\operatorname{Spec} k$ into X is a map into some affine open subscheme $\operatorname{Spec} T$ of X, and such a morphism is determined by a k-algebra map $T \to k$—that is, by a maximal ideal of T with residue class field k. Conversely, we may reverse the construction and see that any k-rational closed point p gives rise to a unique morphism $\operatorname{Spec} k \to X$.

It seems appropriate to reiterate here the warning about working in the category of S-schemes rather than the category of all schemes, where applicable. Consider, for example, the following exercise.

EXERCISE IV-7 ____

Let $X = \operatorname{Spec} \mathbf{C}$, considered as an abstract scheme—that is, a scheme over $\mathbf{Z}$. Describe the set $h_X(\operatorname{Spec} \mathbf{C})$ of all $\mathbf{C}$-valued points of $\operatorname{Spec} \mathbf{C}$.

iv Tangent Spaces to a Functor

Sometimes it is much easier to compute geometric information about a scheme if one knows its functor of points than if one knows its equations! A typical example occurs with the Zariski tangent space. (We will see this applied in Section IV-B-vi.)

We will work with schemes over a fixed field k, and all morphisms will be morphisms over k.

Recall that if X is a scheme, then for any point $p \in X$ the Zariski tangent space to X at p is $\mathrm{Hom}_\kappa(\mathfrak{m}/\mathfrak{m}^2, \kappa)$, where $\mathfrak{m} = \mathfrak{m}_{X,p}$ is the maximal ideal in the local ring of X at p and $\kappa = \kappa(p) = \mathcal{O}_{X,p}/\mathfrak{m}_{X,p}$ is the residue field of X at p. Now let X be a scheme over k. Let Y be the affine scheme $\mathrm{Spec}\, k[\varepsilon]/(\varepsilon)^2$. We claim that *a $k[\varepsilon]/(\varepsilon)^2$-valued point of X is the same as a k-rational closed point p of X together with an element of the Zariski tangent space to X at p.*

To check this assertion, note first that because of the inclusion map $\iota : \mathrm{Spec}\, k \to \mathrm{Spec}\, k[\varepsilon]/(\varepsilon)^2$ induced by the algebra map $k[\varepsilon]/(\varepsilon)^2 \to k$, a morphism $\mathrm{Spec}\, k[\varepsilon]/(\varepsilon)^2 \to X$ determines a morphism $\mathrm{Spec}\, k \to X$—and thus a closed, k-rational point p of X. The data of an extension of such a morphism $\mathrm{Spec}\, k \to X$ is exactly a lifting of the k-algebra homomorphism $\mathcal{O}_{X,p} \to \kappa(p) = k$ to a (local) homomorphism $\mathcal{O}_{X,p} \to k[\varepsilon]/(\varepsilon)^2$. Such a homomorphism induces a map from the maximal ideal $\mathfrak{m}_{X,p}$ to the maximal ideal (ε) of $k[\varepsilon]/(\varepsilon)^2$; and since $(\varepsilon)^2 = 0$, this map factors through a map $t : \mathfrak{m}_{X,p}/\mathfrak{m}_{X,p}^2 \to (\varepsilon) \cong k$. The map t is an element of $(\mathfrak{m}_{X,p}/\mathfrak{m}_{X,p}^2)^*$, the Zariski tangent space to X at p.

Conversely, given p and $t : \mathfrak{m}_{X,p}/\mathfrak{m}_{X,p}^2 \to (\varepsilon) \cong k$, we may construct a map $\mathcal{O}_{X,p} \to k[\varepsilon]/(\varepsilon)^2$ from the map $\pi : \mathcal{O}_{X,p} \to \kappa(p) = k$ corresponding to p: π and the k-algebra structure map $k \to \mathcal{O}_{X,p}$ define a k-vector space splitting of $\mathcal{O}_{X,p}/\mathfrak{m}_{X,p}^2$ into $k \oplus \mathfrak{m}_{X,p}/\mathfrak{m}_{X,p}^2$; and we use the identity map on k and the map t on $\mathfrak{m}_{X,p}/\mathfrak{m}_{X,p}^2$ to define a map $\mathcal{O}_{X,p}/\mathfrak{m}_{X,p}^2 \to k[\varepsilon]/(\varepsilon)^2$, which by composition with the projection gives a map $\mathcal{O}_{X,p} \to k[\varepsilon]/(\varepsilon)^2$. This map determines the desired morphism of schemes $\mathrm{Spec}\, k[\varepsilon]/(\varepsilon)^2 \to X$.

In Section IV-B-ii we will see how this characterization of tangent vectors to a scheme may be used to give an intrinsic description of tangent vectors to projective space.

If now F is a functor from the category of k-algebras to (sets) and $p \in F(k)$, we may define the tangent space $T_p F$ to F at p to be the fiber over p in $F(k[\varepsilon]/\varepsilon^2) \to F(k)$. One objection that may be raised to this definition is that it gives us the tangent space as a set, rather than as a vector space over k. One can at least define multiplication by elements of k: if $a \in k$, then the map $\varepsilon \mapsto a\varepsilon$ induces an endomorphism of the algebra $k[\varepsilon]/\varepsilon^2$, making the diagram

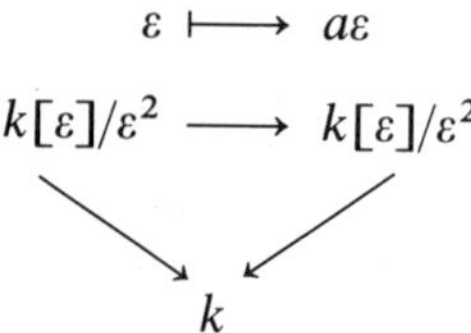

commute; and this induces a map $T_p F \to T_p F$, which is the desired multiplication by a. To make $T_p F$ a vector space, we now need to define an addition map $T_p F \times T_p F \to T_p F$. In general, there seems to be no way to do this; but for those functors F that, like representable functors, "preserve limits," we can.

To do this, consider the scheme $\operatorname{Spec} k[\varepsilon', \varepsilon'']/(\varepsilon', \varepsilon'')^2$—that is, the closed subscheme of the plane given by the square of the maximal ideal of a point. The commutative diagram

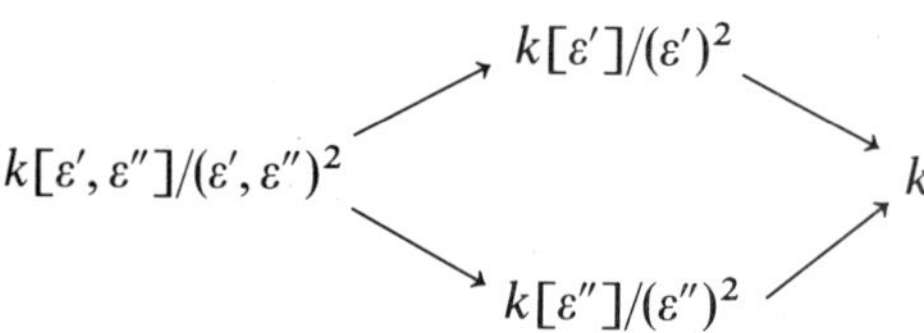

given by the projections expresses $k[\varepsilon', \varepsilon'']/(\varepsilon', \varepsilon'')^2$ as the fibered product, over k, of $k[\varepsilon']/(\varepsilon')^2$ and $k[\varepsilon'']/(\varepsilon'')^2$. On the other hand, there is a third map, $\sigma : k[\varepsilon', \varepsilon'']/(\varepsilon', \varepsilon'')^2 \to k[\varepsilon]/(\varepsilon)^2$, that takes *both* ε' and ε'' to ε. Thus if the functor F preserves fibered products in the category of algebras, or even just the fibered product given above, we can form the diagram

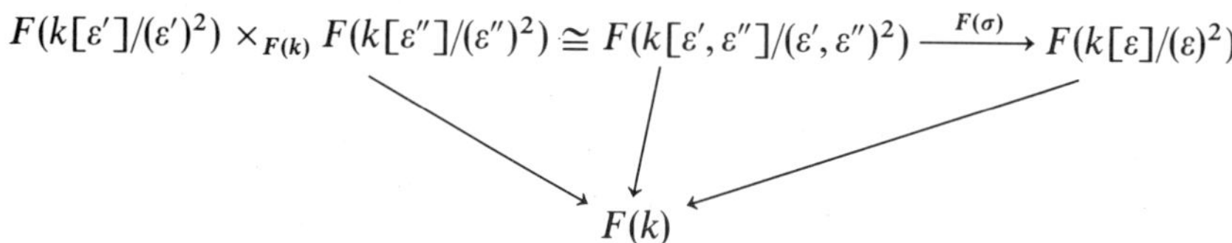

and taking fibers over $p \in F(k)$, we get the desired addition map.

EXERCISE IV-8

Verify that these maps make $T_p F$ into a k-vector space, and that this is the old vector space structure in the case where F is a representable functor.

v Group Schemes

It is extremely easy to specify extra structure on a scheme by specifying it on the functor. For example, we may define a group scheme as follows.

A *group scheme* is a scheme G and a factorization of the functor $h_G : (\text{ring}) \to (\text{sets})$ through the "forgetful functor" $(\text{groups}) \to (\text{sets})$. That is, a group scheme is a scheme and a natural way of regarding $\text{Morph}(X, G)$ as a group for each X.

By Yoneda's lemma IV-1 this is the same thing as giving maps

$$G \times G \to G \qquad G \to G \qquad \text{Spec}\, k \to G$$

representing the multiplication, inverse, and identity element, respectively, and satisfying the usual laws (associativity and so on), but is often much simpler. For example, GL_n can be defined as the affine scheme of invertible integral $n \times n$ matrices,

$$\text{Spec}\, \mathbf{Z}[x_{ij}]\,[\det(x_{ij})^{-1}]$$

but one usually thinks of it as a functor that associates to every ring T the group $GL_n(T)$. The interesting point here is just that this family of groups already specifies the structure of a scheme and the additional structure maps!

B Characterization of a Space by Its Functor of Points

It is often much easier—and sometimes more illuminating—to describe the functor of points of an interesting scheme than to prove the existence theorem that there is a scheme of which it is the functor of points. Typical examples come from the Hilbert scheme and other moduli problems, where one wants a natural space whose points represent some geometric objects. However, the point of view is also useful in discussing much simpler objects, such as fibered products, or projective space itself. We start with these.

i Fibered Products

The construction of fibered products in the category of schemes, which was explained in Chapter I, is of fundamental importance, but it is surprisingly clumsy. Using the functorial point of view, we can at least describe the fibered product of two schemes by giving its functor of points directly.

Recall that if A, B, and C are functors from some category $\mathscr{C}$ to the category of sets, and if $f : A \to C$ and $g : B \to C$ are morphisms of functors, then the fiber product $A \times_C B$ is the functor from $\mathscr{C}$ to (sets) defined, for any object Z of $\mathscr{C}$, by

$$A \times_C B(Z) = \{(a,b) \in A(Z) \times B(Z) \mid f(Z)(a) = f(Z)(b) \text{ in } C(Z)\}$$

and defined on morphisms of $\mathscr{C}$ in the obvious way.

If $X \to S$ and $Y \to S$ are morphisms of schemes, then the fibered product $X \times_S Y$ is determined by saying that it is the scheme whose functor of points is the fibered product of functors $h_X \times_{h_S} h_Y$.

The next example gives a different way of looking at maps to projective spaces.

ii Projective Space

Theorem III-13 may be translated immediately into our new language.

THEOREM IV-9   If $Y = \mathbf{P}_k^n$, then

$$h_Y(X) = \{\text{locally free subsheaves } K \subset \mathcal{O}_X^{n+1} \text{ that locally are summands of rank } n\}$$

$$= \frac{\{\text{invertible sheaves } P \text{ on } X \text{ with an epimorphism } \mathcal{O}_X^{n+1} \to P\}}{\{\text{units of } \mathcal{O}_X(X) \text{ acting as automorphisms of } P\}}$$

By way of an example, let us combine this description of maps to projective space with the earlier characterization of tangent vectors to a scheme X as maps of $\operatorname{Spec} k[\varepsilon]/(\varepsilon)^2$ to X (IV-A-iv) to compute the Zariski tangent spaces to $\mathbf{P}_k^n$ at k-valued points. By what we have just said, a k-valued point of $\mathbf{P}_k^n$ is a rank n summand $K \subset k^{n+1}$; let L be the quotient, $L = k^{n+1}/K$. The Zariski tangent space T at this point is the set of all summands $K' \subset (k[\varepsilon]/(\varepsilon)^2)^{n+1}$ that restrict to K modulo ε. We claim that there is a natural isomorphism

$$T \cong \operatorname{Hom}_k(K, L)$$

To produce it, choose a splitting $k^{r+1} = K \oplus L$ and a basis e_i of K. Any summand K' of $(k[\varepsilon]/(\varepsilon)^2)^{r+1}$ that reduces mod ε to K has a basis of the form $\{e_i + \varepsilon s_i + \varepsilon t_i\}$ with s_i in K and t_i in L, say. We associate to K' the map $\alpha : K \to L$ sending e_i to t_i. Conversely, given any map α, we may define K' to be the module spanned by the elements $e_i + \varepsilon\alpha(e_i)$.

EXERCISE IV-10 ____

Check that these definitions are independent of all the choices made.

iii Characterization of Schemes Among Functors

As we have said, one of the principal reasons for introducing the functor of points is that it embeds the category of schemes in the larger category of functors. In this larger category many constructions, such as fibered products, are more readily made; and as we will illustrate in our discussion of the Hilbert scheme in the following section, it is often easier to describe a scheme as a functor than to give its equations.

To the extent that we want to define and/or construct schemes first as functors, we run into a fundamental problem: the problem of determining when a functor comes from a scheme. Here is one characterization.

We say that $F : (\text{rings}) \to (\text{sets})$ is a *sheaf in the Zariski topology* if for each ring R and each open covering of $X = \operatorname{Spec} R$ by distinguished open affines $U_i = \operatorname{Spec} R_{f_i}$ the functor F satisfies the analogue of the sheaf axiom for the "open covering" $\bigcup h_{U_i} = h_X$. That is, for every collection of elements $\alpha_i \in F(R_{f_i})$ such that α_i and α_j map to the same element in $F(R_{f_i f_j})$, there is a unique element $\alpha \in F(R)$ mapping to each of the α_i.

This is a property that is reasonably easy to check in practice. It is in fact enough to guarantee that F comes from a scheme if we know already that F is covered by affine schemes in the following sense. The reader may prove the following theorem.

THEOREM IV-11   A functor $F : (\text{rings}) \to (\text{sets})$ is of the form h_Y for some scheme Y iff

1. F is a sheaf in the Zariski topology,
2. there exist rings R_i and elements $\alpha_i \in F(R_i)$—that is, by IV-1, maps

$$\alpha_i : h_{R_i} \to F$$

—such that for every field k, $F(k)$ is the union of the images of $h_{R_i}(k)$ under the maps α_i.

As an easy application, one can use the theorem to show the existence of fibered products.

EXERCISE IV-12 ____

a. Show that if $f : A \to C$ and $g : B \to C$ are morphisms of functors all of which are sheaves in the Zariski topology, then the fibered product $A \times_C B$ is a sheaf in the Zariski topology.
b. Use the open covering of the fibered product suggested in Chapter I and the above theorem to prove the existence of fibered products in the category of schemes.

The theorem can also be used to prove the existence of the Grassmannian scheme from its functorial description.

EXERCISE IV-13 ——————————————————————————

For $0 < m < n$, let

$$G = G_{n,m} : (k\text{-algebras}) \to (\text{sets})$$

be the functor given by

$$G(T) = \{\text{rank } m \text{ direct summands of } T^n\}$$

Prove that this is the functor of points of a closed subscheme of projective space, called the *Grassmannian* of m-planes in n-space, as follows.

First, let $r = \binom{n}{m} - 1$, and define a map $G \to h_{\mathbf{P}^r}$ by sending a summand $M \subset T^n$ to $\bigwedge^m M \subset \bigwedge^m T^n$. Cover $\mathbf{P}^r$ by the usual open affine subschemes; show that these are represented by the subfunctors

$$U_i(T) = \{\text{rank } r \text{ summands of } T^{r+1} \text{ such that the } i\text{th basis vector}$$
$$\text{of } T^{r+1} \text{ generates the cokernel}\}$$

(Note that we do *not* have $h_{\mathbf{P}^r}(T) = \bigcup U_i(T)$, as one might naively expect, except on local rings.) Check that the intersection of U_i and G represents an affine scheme. Show that G is a sheaf in the Zariski topology, and conclude that G represents a scheme.

We will see other examples of the use of this theorem in the following section. For more on the theory, the functorial point of view is developed in far greater depth and detail in the book of Demazure and Gabriel (1970), to which the interested reader is referred for more information.

One of the principal goals in Grothendieck's work on schemes was to find a characterization of scheme-functors by weak general properties that could often be checked in practice and so lead to many existence theorems in algebraic geometry (like Brown's theorem in [Hot]; see Spanier [1966, Chap. 7.7]). It seemed at first that this program would fail completely and that scheme-functors were really quite special (see, for instance, Hironaka [1962] and Mumford [1962, p. 83]). Artin, however, discovered an extraordinary approximation theorem exhibiting a category of functors $\mathscr{F}$ only a "little" larger than the scheme-functors that can indeed be characterized by weak general properties. Geometrically speaking, the functors $\mathscr{F}$ are like spaces obtained by dividing affines by "étale equivalence relations" and then gluing. He calls these *algebraic spaces* (for a typical occasion where they arise, see Section IV-B-vii). For details, see Artin (1971) and Knutson (1971).

iv The Hilbert Scheme

Probably the one area where the notion of the functor of points has had the most impact is in the construction and description of parameter spaces. We have already mentioned in Section III-C-iii, for example, that the subschemes of a projective space $\mathbf{P}^n$ over a field k having a given Hilbert polynomial P form a scheme, which we will call $\mathscr{H}_P$. It seems rather surprising that such a statement has an unambiguous meaning. While it is intuitively plausible that the *set* of such objects might form the points of a variety, couldn't they form a variety in several different ways? And in what sense do they form a scheme?

The answers are obtained by making precise what properties we want the correspondence between the set of subschemes and the set of points of $\mathscr{H}_P$ to have. Specifically, note that if $\mathscr{X} \subset \mathbf{P}^n \times B \to B$ is any flat family of subschemes of $\mathbf{P}^n$ with Hilbert polynomial P, we get a map from the points of B with residue field k to the points of $\mathscr{H}_P$ with residue field k, sending a point $b \in B$ to the point of $\mathscr{H}_P$ corresponding to the fiber X_b of $\mathscr{X}$ over b. It is natural to ask that this map come from a regular map $B \to \mathscr{H}_P$. Carrying this a little further, we want $\mathscr{H}_P$ to have the property that for any scheme B over k, the set of flat families of subscheme of $\mathbf{P}^n$ with Hilbert polynomial P parametrized by B is naturally identified with the set of maps from B to $\mathscr{H}_P$.

To say this a little differently, we define the *functor h_P of flat families of schemes in $\mathbf{P}^n$ with Hilbert polynomial P* to be the functor

$$h_P : (\text{schemes over } k)^o \to (\text{sets})$$

that associates to any B the set of subschemes $\mathscr{X} \subset \mathbf{P}^n \times B$ flat over B whose fibers over points of B have Hilbert polynomial P. We then want $\mathscr{H}_P$ to be the scheme that represents h_P—in other words, the scheme whose functor of points is h_P. By Yoneda's lemma IV-1, this determines the scheme $\mathscr{H}_P$, if one such exists; the key theorem is thus the following.

THEOREM IV-14   There exists a scheme $\mathscr{H}_P$ whose functor of points is the functor h_P.

In fact, it turns out that $\mathscr{H}_P$ is often a scheme that is genuinely *not* a variety. This is one of the ways in which the theory of schemes arises even if one's original purpose is simply to study varieties. We will describe, in Exercises IV-25–IV-27 below, a famous example of Mumford (1962) of a Hilbert scheme that is nonreduced even at points corresponding to smooth, irreducible projective varieties.

We should remark also that the definition of the Hilbert scheme $\mathscr{H}_P$ may be generalized in many ways. To mention one, suppose that $Y \subset \mathbf{P}^n$ is a closed subscheme. It is then the case that the functor h_P^Y associating to any scheme

B the set of flat families $\mathscr{X} \subset Y \times B \subset \mathbf{P}^n \times B$ of schemes with Hilbert polynomial P is a closed subfunctor of the functor h_P, and so is represented by a closed subscheme $\mathscr{H}_P^Y \subset \mathscr{H}_P$.

EXERCISE IV-15

Verify both parts of the last assertion.

The statement of Theorem IV-14 has another interpretation that is often useful: saying that the functor h_P is representable is the same thing as saying that there exists a *universal family*—that is, a scheme $\mathscr{H}$ and a subscheme $\mathscr{X} \subset \mathbf{P}^n \times \mathscr{H}$ flat over $\mathscr{H}$ with Hilbert polynomial P—such that any subscheme $Y \subset \mathbf{P}^n \times B$ flat over B with Hilbert polynomial P is equal to the fiber product $Y = \mathscr{X} \times_{\mathscr{H}} B \subset \mathbf{P}^n \times B$ for a unique morphism $B \to \mathscr{H}$. Clearly, if a universal family $\mathscr{X} \subset \mathbf{P}^n \times \mathscr{H}$ exists, then $\mathscr{H}$ represents the functor h_P. Conversely, if a scheme $\mathscr{H}$ represents h_P, then the subscheme $\mathscr{X} \subset \mathbf{P}^n \times \mathscr{H}$ associated to the identity map is universal in the above sense.

We will not give a proof of Theorem IV-14 but will indicate how it may be approached; for more details we refer the reader to Mumford (1966) or Grothendieck's original Seminaire Bourbaki talk (1957–1962).

The idea is easy to summarize: it is that the subscheme $X \subset \mathbf{P}^r$ is determined by its ideal $I(X)$, which in turn is determined for m sufficiently large (in a sense depending only on P) by its degree m piece $I(X)_m \subset S_m$. This in turn corresponds to a point in the Grassmannian of planes of codimension $P(m)$ in the space S_m. In this way $\mathscr{H}_P$ becomes a subscheme of the Grassmannian of such planes.

The key point is that we can choose a single m that has this property uniformly for every subscheme X with Hilbert polynomial P; that is, that for every P, there is an m_0 such that if $m \geq m_0$ and X is a subscheme of $\mathbf{P}^n$ with Hilbert polynomial P, then $I(X)_{l \geq m}$ is generated by $I(X)_m$ and the codimension of $(I(X)_m)$ in S_m is $P_X(m)$. Examining our proof that the Hilbert polynomial is a polynomial, we see that to prove this it is enough to show that the degrees of the generators of the free modules in the minimal free resolution of $I(X)$ can be bounded in terms of the Hilbert polynomial of X. This is done by using the idea of "Castelnuovo regularity" of X; a more complete description would take us too far afield.

Given this, suppose we have a flat family $\mathscr{X} \subset \mathbf{P}^n \times B$. For each k-valued point in $\mathscr{X}$ we get, as above, a point in the Grassmannian G. With a little more technique than we now have at hand, it is not hard to see that this association extends to an actual map of schemes $\varphi : B \to G$.

To finish the argument, one must show that there exists a subscheme $\mathscr{H}_P \subset G$ such that a morphism $\varphi : B \to G$ comes in this way from a flat family with Hilbert polynomial P if and only if φ factors through $\mathscr{H}_P$. We will content ourselves here with describing the equations of $\mathscr{H}_P$ as a closed subscheme of G. Let $\mathscr{S}$ be the universal subbundle on G. We have multiplication maps

$$\mathrm{mult}_k : \mathscr{S} \otimes S_k \to S_{k+m}$$

and we can take $\mathscr{H}_P$ to be the "determinantal" subscheme of G defined by the conditions that

$$\mathrm{rank}(\mathrm{mult}_k) \leq \dim S_{k+m} - P(m+k)$$

for all $k \geq 0$. It is immediate that the desired maps φ all factor through this subscheme. Given any point p this subscheme, one shows that (because m has been chosen so large) the ideal generated by $\mathscr{S}_p \subset S_m$ defines a scheme with Hilbert polynomial P. This gives us a "tautological family" on $\mathscr{H}_p$ of schemes with Hilbert polynomial P. Given any map φ from a scheme B into $\mathscr{H}_p$, one can "pull back" this family by using the fibered product to get a family over B, and the map φ will be associated to this family. Once all this is verified, the description of $\mathscr{H}_P$ by its functor of points ensures that $\mathscr{H}_P$ does not depend on the choice of m.

v Examples of Hilbert Schemes

We will mention, largely in the form of exercises, some examples of Hilbert schemes. To begin with, the Grassmannians themselves are examples of Hilbert schemes: they parametrize subschemes of degree 1 in $\mathbf{P}^n$. The following exercise deals with the simplest special case of this, but the general statement (and proof) differs only numerically.

EXERCISE IV-16 ──

Let $P(m) = m + 1$ be the Hilbert polynomial of a line. Show that the Hilbert scheme of subschemes of $\mathbf{P}^3$ with Hilbert polynomial P is the Grassmannian introduced in Exercise IV-13.

Note that if $Y \subset \mathbf{P}^n$ is a closed subscheme, the functor of flat families of k-planes contained in Y is a closed subfunctor of the functor of k-planes in $\mathbf{P}^n$, and so is represented by a closed subscheme $F_k(Y)$ of the Grassmannian. This subscheme $F_k(Y)$ is called the *Fano scheme* of Y.

Another very straightforward example is hypersurfaces. It is a standard observation that the set of hypersurfaces of degree d in $\mathbf{P}^n$ may be identified with the points of the projective space of homogeneous polynomials of degree d in $n + 1$ variables. In fact, this projective space turns out to be the Hilbert scheme of such hypersurfaces. As in the preceding example, the following exercise deals with one typical example.

EXERCISE IV-17 ____

Let $P(m) = 2m + 1$ be the Hilbert polynomial of a conic curve in $\mathbf{P}^2$. Show that the Hilbert scheme $\mathcal{H}_P$ of subschemes of $\mathbf{P}^2$ with Hilbert polynomial P is $\mathbf{P}^5$.

Beyond these examples, the geometry of Hilbert schemes is much less well known. Even the Hilbert schemes parametrizing zero-dimensional subschemes of $\mathbf{P}^n$ remain mysterious: Iarrobino (1985), for example, has shown (contrary to naive expectations) that these are not in general irreducible. In the case of $\mathbf{P}^2$, they are in fact irreducible and smooth, but their global geometry presents many problems: see, for example, Collino (1988). One case where we can actually give a description is the following exercise.

EXERCISE IV-18 ____

(For those with a background in projective geometry.) Let P be the constant polynomial 2. Show that the Hilbert scheme $\mathcal{H}_P$ parametrizing subschemes of $\mathbf{P}^2$ with Hilbert polynomial P may be obtained by blowing up the product $\mathbf{P}^2 \times \mathbf{P}^2$ along the diagonal and then taking the quotient by the involution-exchanging factors.

In the general setting our knowledge of Hilbert schemes is minimal. For example, in the case of curves in $\mathbf{P}^3$—the simplest example of a Hilbert scheme parametrizing schemes that are pure positive-dimensional but not hypersurfaces—we do not have even a guess as to the number of components of $\mathcal{H}_P$, their dimension, or their smoothness or singularity (for a discussion of this case, see Harris and Eisenbud [1982]).

vi Tangent Spaces to Hilbert Schemes

One facet of the Hilbert scheme that is best described in terms of its functor of points is its tangent space at a point. We first introduce the notion of a first-order deformation: if Y is any scheme and $X \subset Y$ a closed subscheme, a *first-order deformation of X in Y* is defined to be a flat family $\mathcal{X} \subset Y \times \operatorname{Spec} k[\varepsilon]/(\varepsilon^2)$ such that the fiber of $\mathcal{X}$ over the reduced point $\operatorname{Spec} k \subset \operatorname{Spec} k[\varepsilon]/(\varepsilon^2)$ is X. It then follows, via the characterization of tangent vectors to schemes given in IV-A-iv, that *the tangent space to the Hilbert scheme $\mathcal{H}_P$ at a point $[X]$ is the space of first-order deformations in $\mathbf{P}^n$ of X*; and more generally, if $Y \subset \mathbf{P}^n$ is a projective scheme, then the tangent space to the Hilbert scheme $\mathcal{H}_P^Y$ at a point $[X]$ is the space of first-order deformations of X in Y.

This is especially useful since the space of first-order deformations may often be calculated, even in circumstances where we have no hope of writing down the equations of $\mathscr{H}_P$. To do this, we introduce the *normal sheaf* $\mathscr{N}_{X/Y}$ to a closed subscheme X of a scheme Y: this is defined to be the sheaf

$$\mathscr{N}_{X/Y} = \operatorname{Hom}_{\mathscr{O}_X}(\mathscr{I}/\mathscr{I}^2, \mathscr{O}_X) = \operatorname{Hom}_{\mathscr{O}_Y}(\mathscr{I}, \mathscr{O}_X)$$

where $\mathscr{I} = \mathscr{I}_{X/Y}$ is the ideal sheaf of X in Y. We then have the following basic theorem.

THEOREM IV-19   Given a closed subscheme X of a scheme Y, the space of first-order deformations of X in Y is the space of global sections of its normal sheaf $\mathscr{N}_{X/Y}$.

Proof   To begin with, let $\mathscr{X} \subset Y \times \operatorname{Spec} k[\varepsilon]/(\varepsilon^2)$ be any subscheme whose intersection with the fiber $Y \cong Y \times \operatorname{Spec} k \subset Y \times \operatorname{Spec} k[\varepsilon]/(\varepsilon^2)$ is X (do not assume $\mathscr{X}$ is flat). Let $U \subset Y$ be any affine open subset, $V = X \cap U$ the corresponding affine open subset of X, and $\mathscr{V} = \mathscr{X} \cap U \times \operatorname{Spec} k[\varepsilon]/(\varepsilon^2)$. Let $A = \mathscr{O}_Y(U)$ be the coordinate ring of U and $I = I(V)$ the ideal of V in A, so that the restriction to V of the sheaf $\mathscr{N}_{X/Y}$ is the sheaf associated to the A-module $\operatorname{Hom}(I, A/I)$.

The coordinate ring of $U \times \operatorname{Spec} k[\varepsilon]/(\varepsilon^2)$ is $A \otimes k[\varepsilon]/(\varepsilon^2)$; we write an element of this ring as $f + \varepsilon g$, with f and $g \in A(U)$ and in particular, we may write the ideal $I(\mathscr{V})$ of $\mathscr{V}$ as

$$I(\mathscr{V}) = (f_1 + \varepsilon g_1, f_2 + \varepsilon g_2, \ldots, f_k + \varepsilon g_k)$$

where by hypothesis the elements $f_i \in A$ generate the ideal I. We claim now that *there exists an A-module homomorphism $\varphi : I \to A/I$ carrying f_i to g_i if and only if $\mathscr{V} \to \operatorname{Spec} k[\varepsilon]/(\varepsilon^2)$ is flat* (note that if φ exists, it is unique). The theorem follows immediately from this claim: in one direction, if the family $\mathscr{X} \to \operatorname{Spec} k[\varepsilon]/(\varepsilon^2)$ is flat, then by uniqueness the homomorphisms φ patch together to give a section of the sheaf $\mathscr{N}_{X/Y}$; while given a global section of $\mathscr{N}_{X/Y}$, we can simply take $\mathscr{X}$ to be given locally by the ideal

$$\{ f + \varepsilon \cdot \varphi(f) : f \in I(V) \}$$

To prove the claim, note first that a $k[\varepsilon]/(\varepsilon^2)$-module M is flat if and only if when we tensor the exact sequence of $k[\varepsilon]/(\varepsilon^2)$-modules

$$0 \to (\varepsilon) \to k[\varepsilon]/(\varepsilon^2) \to k \to 0$$

by M, it remains exact. Applying this to the coordinate ring $B = \mathscr{O}_{\mathscr{X}}(\mathscr{V})$ of $\mathscr{V}$, we see that $\mathscr{V}$ will be flat over $\operatorname{Spec} k[\varepsilon]/(\varepsilon^2)$ if and only if the map

$$(\varepsilon) \otimes B \to B$$

is injective—that is, if and only if, for any $f \in A$,

$$\varepsilon \cdot f \in I(\mathscr{V}) \Rightarrow f \in I(V)$$

Suppose now that $f \in A$ and $\varepsilon \cdot f \in I(\mathscr{V})$. We can then write

$$\varepsilon \cdot f = \sum (a_i + \varepsilon b_i) \cdot (f_i + \varepsilon g_i) = \sum a_i f_i + \varepsilon \cdot \sum (a_i g_i + b_i f_i)$$

We know that the first term on the right is zero, since the other two terms in the equality are divisible by ε. Now, if there exists a homomorphism $\varphi : I \to A/I$ of A-modules such that $\varphi(f_i) = g_i$, then we can write

$$\sum a_i \cdot g_i = \sum a_i \cdot \varphi(f_i) = \varphi\left(\sum a_i f_i\right) = 0$$

so the existence of a module map φ carrying f_i to g_i implies that the family $\mathscr{V}$ is flat.

Conversely, suppose that $\mathscr{V} \to \operatorname{Spec} k[\varepsilon]/(\varepsilon^2)$ is flat. Then for any collection of $a_i \in A$ such that $\sum a_i f_i = 0$, we have

$$\varepsilon \cdot \sum a_i \cdot g_i = \sum a_i \cdot (f_i + \varepsilon g_i) \in I(\mathscr{V}) \Rightarrow \sum a_i \cdot g_i \in I(V)$$

We can thus define an A-module map $\varphi : I \to A/I$ by sending, for any $a_1, \ldots, a_k \in A$, the element $\sum a_i f_i \in I$ to the element $\sum a_i g_i \in A/I$; by the last calculation this will be well defined. $\qquad\square$

One trivial but quite useful consequence of this theorem is the following corollary.

COROLLARY IV-20   The dimension of any irreducible component Σ of the Hilbert scheme is at most the dimension of the space of sections of the normal sheaf of any scheme X with $[X] \in \Sigma$.

In fact, this *a priori* estimate for the dimension of a component of the Hilbert scheme gives the right answer more often than not, especially when applied to a general point $[X]$ of a component of $\mathscr{H}_P$. The following exercises give examples of this.

EXERCISE IV-21 ____

Let Σ be the component of the Hilbert scheme whose general member is a complete intersection $X \subset \mathbf{P}^n$ of k hypersurfaces of degree d. Calculate the dimension of the space of global sections of $\mathcal{N}_X$ and show that this is equal to the dimension of Σ.

EXERCISE IV-22 ____

Generalize the preceding exercise to the case of the component of the Hilbert scheme whose general member is a complete intersection $X \subset \mathbf{P}^n$ of hypersurfaces of degrees $d_1, \ldots, d_k$. (This can get complicated; you may want to stick to the case $k = 2$, which is enough to see how it goes.)

EXERCISE IV-23 ____

Let $P(m)$ be the polynomial $3m + 1$ and let Σ be the component of the Hilbert scheme $\mathcal{H}_P$ of subschemes of $\mathbf{P}^3$ whose general member is a twisted cubic curve C. Show that the dimension of Σ is 12, and that this is equal to the dimension of the space of sections of $\mathcal{N}_C$.

EXERCISE IV-24 ____

By way of warning, the component Σ of the preceding exercise is not the only component of the Hilbert scheme: there is another component Σ' whose general member is the (disjoint) union of a plane cubic curve and a point. These two intersect in the locus of schemes $C \subset \mathbf{P}^3$ such that C is the union of a plane curve C_0 and the double point (that is, the scheme given by the square of the maximal ideal of a point) supported at a singular point of C_0. At a point of their intersection $\mathcal{H}_P$ is singular, and its tangent space will be strictly larger than the dimension of either Σ or Σ'. Verify this.

It is not always the case, however, that the dimension of a component of the Hilbert scheme is equal to the dimension of the space of sections of the normal sheaf of a general member; there are examples of Hilbert schemes that are nonreduced along whole components, even when the general points of those components correspond to smooth, irreducible varieties. The first example of this is due to Mumford (1962); the following series of exercises describes it.

Mumford's example deals with curves of degree 14 and genus 24 in $\mathbf{P}^3$. There are (as we shall see) several components of the Hilbert scheme parametrizing such curves; we will be concerned with the component whose general member lies on a smooth cubic surface. By way of notation, let $P(m) = 14m - 23$ be the Hilbert polynomial of a curve of degree 14 and genus 24, and let $\mathcal{H} = \mathcal{H}_P$ be the Hilbert scheme parametrizing subschemes of $\mathbf{P}^3$ with this Hilbert polynomial. We will denote by Σ the subset of $\mathcal{H}$ corresponding to smooth curves $C \subset \mathbf{P}^3$ of degree 14 and genus 24 that are contained in a smooth cubic surface S and linearly equivalent on S to $4H + 2L$, where H is the hyperplane divisor and L a line on S.

EXERCISE IV-25* ____

Show that Σ is a constructible subset of $\mathcal{H}$ and that its closure $\bar{\Sigma}$ in $\mathcal{H}$ has dimension 56.

Not all curves of degree 14 and genus 24 in $\mathbf{P}^3$ have to lie on cubic surfaces. Thus, it is not *a priori* clear that the subvariety $\bar{\Sigma} \subset \mathcal{H}$ is an irreducible component of $\mathcal{H}$: the curves C parametrized by $\bar{\Sigma}$ could be specializations of other curves not lying on cubics. To see that this is not in fact the case, we make another dimension count.

EXERCISE IV-26* ____

Let C be a smooth, irreducible curve of degree 14 and genus 24 in $\mathbf{P}^3$, and assume that C does not lie on a cubic surface. Show that it must lie on two quartic surfaces T, T' not having a common component, and that the residual intersection of T and T' (that is, the union of the irreducible components of $T \cap T'$ other than C) is a curve of degree 2. By analyzing what this residual intersection may look like, show that the set of such curves C is a constructible subset of $\mathcal{H}$ whose closure has dimension at most 56. Deduce that the subvariety $\bar{\Sigma}$ of Exercise IV-25 is indeed an irreducible component of $\mathcal{H}$.

EXERCISE IV-27 ____

Now let C be a smooth curve of degree 14 and genus 24 lying on a smooth cubic surface $S \subset \mathbf{P}^3$. Using the exact sequence

$$0 \to \mathcal{N}_{C/S} \to \mathcal{N}_{C/\mathbf{P}^3} \to \mathcal{N}_{S/\mathbf{P}^3} \otimes \mathcal{O}_C \to 0$$

(where for any pair of schemes $X \subset Y$ we write $\mathcal{N}_{X/Y}$ for the normal sheaf $\mathrm{Hom}(\mathcal{I}_{X/Y}, \mathcal{O}_X)$ of X in Y), show that the dimension of the space of sections of the normal sheaf $\mathcal{N}_C = \mathcal{N}_{C/\mathbf{P}^3}$ is 57. Deduce that $\mathcal{H}$ is nowhere reduced along $\bar{\Sigma}$.

vii Moduli Spaces

A similar situation arises when we want to construct a moduli space of geometric objects. For example, we would like to identify the set of non-singular, projective curves of genus g over a field k with the set of closed points of a "moduli scheme" $\mathcal{M}_g$. To avoid unnecessary complication, we restrict to the case $\operatorname{char}(k) = 0$. Again, the way to express what we want is to introduce the *functor of nonsingular curves of genus g*: this is the functor

$$\mathcal{M}_g^{\mathrm{fun}} : (\text{schemes over } k)^o \to (\text{sets})$$

that assigns to any scheme B over k the set of flat morphisms $\pi : \mathcal{X} \to B$ whose fibers are nonsingular curves of genus g. We define $\tilde{\mathcal{M}}_g$ to be the scheme (if any) that represents the functor $\mathcal{M}_g^{\mathrm{fun}}$.

Since schemes are uniquely determined by their functors of points, the only difficulty with the "definitions" above is whether such schemes exist— that is, whether the given functors are representable. The answer in the case of $\mathcal{H}_P$ is "yes"; and indeed, as we have pointed out, the characterization of $\mathcal{H}_P$ as the scheme that represents the functor h_P is crucial to the proof of existence.

However, in the case of $\tilde{\mathcal{M}}_g$, the answer is "no"! For example, even in the case of smooth curves of genus 1 over $\mathbf{C}$ there does not exist a moduli space $\tilde{\mathcal{M}}_1$ in this sense. To see this, it suffices to show that there does not exist a universal family—in other words, a flat morphism $\pi : \mathcal{C} \to \mathcal{M}$ of schemes with fiber smooth curves of genus 1 such that for every family $\mathcal{Y} \to B$ of smooth curves of genus 1 there are unique maps $\varphi : B \to \mathcal{M}$ and $\Phi : \mathcal{Y} \to \mathcal{C}$ forming a fiber square:

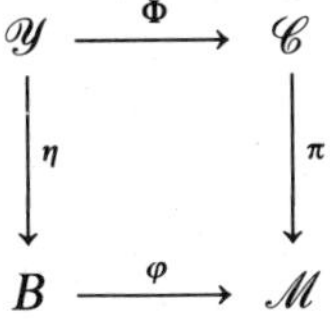

In fact, there does not even exist a tautological family—that is, a morphism $\mathcal{C} \to \mathcal{M}$ and a bijection between the closed points of $\mathcal{M}$ and the set of iso-morphism classes of smooth curves of genus 1 such that the fiber over each point $p \in \mathcal{M}$ is in the isomorphism class corresponding to the point p. We will exhibit two kinds of obstructions to the existence of a universal family, one local and one global.

For the local obstruction, recall that curves of genus 1 over $\mathbf{C}$ are classified by their *j-invariant*: we can write any such curve as the plane cubic

$$y^2 = x(x - 1)(x - \lambda)$$

for some $\lambda \neq 0, 1 \in k$; and two such curves C_λ and $C_{\lambda'}$ will be isomorphic if and only if their *j*-invariants $j(\lambda)$ and $j(\lambda')$ are equal, where

$$j(\lambda) = 256 \cdot \frac{(\lambda^2 - \lambda + 1)^3}{\lambda^2 \cdot (\lambda - 1)^2}$$

It can be shown (see Proposition 1.7 of Silverman [1986], for example) that given a family $\mathscr{X} \to B$ of smooth curves of genus 1 with smooth base B, the function j is a regular function on B; and locally around any point $b \in B$, λ can be defined as a regular function, too (though it is not unique). It follows that if there did exist a tautological family, there would have to exist one with base the affine line $\mathbf{A}^1$ with coordinate j. But no such family can exist, because at the point $j = 0$ the function $\lambda^2 - \lambda + 1$ would vanish, and thus j would vanish triply. Less obviously, because $j'(-1) = 0$, it also follows that j can only assume the value 1728 with even multiplicity. (Note that the values $j = 0$ and 1728 correspond to the elliptic curves with "extra automorphisms"— that is, whose automorphism groups contain the automorphism group of a general elliptic curve as subgroups of index 3 and 2, respectively.)

Turning to the global obstruction, even if a tautological family exists over a variety $\mathscr{M}$ whose points correspond to isomorphism classes of curves, such a family may not be universal; that is, it may not induce a bijection between families over a base B and maps of B to $\mathscr{M}$. For example, if we simply exclude the curves of j-invariant 0 and 1728—that is, consider the functor of families of smooth curves of genus 1 not isomorphic to C_0 or C_{1728}—we might hope that the punctured j-line $\mathscr{M} = \mathbf{A}^1 - \{0, 1728\}$ would be a moduli space; and indeed, a tautological family does exist over this open subset of $\mathbf{A}^1$. It is not universal, however: for example, for any fixed λ, let B' be any variety with fixed-point-free involution τ, and consider the family over $B = B'/\{\tau\}$ formed by taking the quotient of the product $E \times B$ by the involution

$$\iota : ((x, y), p) \mapsto ((x, -y), \tau(p))$$

This is a family all of whose fibers are isomorphic to C_λ, and so it can only come from the constant map $B \to \mathscr{M}$; but it can be shown that the family itself is not trivial.

Note that it is the presence of automorphisms in C_λ that makes this example work. Indeed, an analogous argument shows that we can never have a moduli space for schemes modulo isomorphism when some of the objects to be parametrized admit automorphisms. This explains also the discrepancy between the notions of "tautological" and "universal" family: in the case of the Hilbert scheme $\mathscr{H}$ parametrizing subschemes of $\mathbf{P}^n$, if two families $\mathscr{X}$, $\mathscr{X}' \subset \mathbf{P}^n \times B$ over a variety B correspond to the same map $B \to \mathscr{H}$, it follows that they are equal fiber by fiber and hence equal. By contrast, in the case of a moduli space, it would follow only that they are isomorphic fiber by fiber; if the fibers admitted automorphisms, those isomorphisms would not be unique and so might not fit together to give an isomorphism $\mathscr{X} \cong \mathscr{X}'$.

How do we deal with these difficulties? The most naive (and least satisfactory) way is simply to exclude all schemes with automorphisms from consideration when trying to construct a moduli space. This works in some contexts: for example, since the family of curves of genus g with automorphisms has in a suitable sense codimension $g - 2$ among all curves, if we are concerned in particular with the divisor theory of the moduli space of curves of genus $g \geq 4$, we can afford to look just at the moduli space $\mathscr{M}_g^0$ of automorphism-free smooth curves of genus g, which does exist.

There are two more serious approaches, both of which are in active use. The first is to take $\mathscr{M}_g$ to be the scheme whose functor of points "most closely approximates" $\mathscr{M}_g^{\text{fun}}$. It turns out that there is such a thing, called a *coarse moduli space*, and that it has nice properties; for example, the value of its functor of points at an algebraically closed field k is really the set of isomorphism classes of nonsingular curves over k. The second way out is to enlarge the category of schemes in a different way, to the category of algebraic stacks. A discussion of this would take us too far; we refer the reader to Vistoli (1989, Appendix) for a short treatment and to Mumford (1963) for an introduction to the functorial point of view on moduli spaces.

Hints to Selected Exercises

Chapter I

Hint I-18. Answer: The ring of rational functions is the localization of R at the multiplicatively closed set that is the complement of the union of the minimal primes of R; this may also be described as the direct product of the localizations of R at the minimal primes.

Hint I-24. Reduce to the case X affine and use quasi-compactness.

Hint I-31. Show that $\mathcal{O}_U(U) = k[x, y]$, so $\operatorname{Spec} \mathcal{O}_U(U) \neq U$.

Hint I-44. Factor out a prime of $\mathbf{Z}$, and use the principal ideal theorem and dimension theory for affine rings over a field.

Hint I-45. It is enough to do the affine case.

Chapter II

Hint II-5. $[(x^2 + y^2 - 1)]$ has in its closure both the closed points (λ, μ) corresponding to points on the unit circle in $\mathbf{R}^2$ and the points corresponding to complex conjugate pairs of points (z, w) and $(\bar{z}, \bar{w})$ with $z^2 + w^2 - 1$; the point $(x^2 + y^2 + 1)$, of course, has in its closure only the latter kind.

Hint II-9. Consider the successive quotients $\mathfrak{m}^\alpha/\mathfrak{m}^{\alpha+1}$, where $\mathfrak{m}$ is the ideal of $k[x_1, \ldots, x_n]$ generated by $x_1, \ldots, x_n$, and use Nakayama's lemma.

Hint II-14. There are three types in degree 4, isomorphic to $k[x]/(x^4)$, $k[x, y]/(x^2, y^2)$, and $k[x, y]/(x^2, xy, y^3)$, and four types in degree 5, isomorphic to $k[x]/(x^5)$, $k[x, y]/(xy, x^3, y^3)$, $k[x, y]/(x^2, xy^2, x^2y, y^3)$, and $k[x, y]/(x^2, xy, y^4)$.

Hint II-16. The schemes of the former family have ideals containing three general quadrics. To find a continuously varying invariant associated with the three-dimensional space of quadratic forms in four variables that they generate, consider the lines in this three-dimensional space as points of a two-dimensional projective space. The locus of points in this space corresponding to rank 3 quadrics is defined by a quartic equation (the determinant of the matrix of coefficients of these quadrics) and thus forms a curve of degree 4 in the projective plane. The isomorphism class of this curve is determined by the degree 7 scheme. In general, the curve is smooth, and infinitely many nonisomorphic curves arise in this way.

The second family has ideals generated by pairs of general cubic polynomials in two variables. A two-dimensional vector space of cubics defines a map of degree 3 from the projective line to itself, and Hurwitz's theorem shows that such a map must have exactly four branch points (with multiplicity). The isomorphism classes of sets of four points in the projective line are given by the "j-invariant" of the set of four points, which again can take on any value for the sets of branch points of the maps of degree 3, so that again we get a continuously varying family of examples.

Hint II-18. $\dim_k(y)/(xy, y^2) = 1$.

Hint II-19. Use the Zariski tangent space.

Hint II-22. To show that flat modules are torsion-free, use inclusions of the form $(a) \to R$, for $a \in R$. For the converse, first reduce to the local case. Then show that any flat ideal I is principal, for example, by considering the map $I \otimes I \to I^2$.

Hint II-23. If M is torsion-free, then $\operatorname{Tor}_1^R(M, k)$ is zero, so the A-free resolution of M reduces mod t to the $\overline{A}$-free resolution. Prove the converse by playing directly with elements, using the fact that F_0 is R-torsion-free.

Hint II-29. Using the fact that the Zariski tangent space to X is never more than two-dimensional, show that if such a family existed, there would be one whose total space was a nonsingular surface.

Hint II-31. Begin by splitting the k-algebra map

$$\mathcal{O}_X(X) \to \mathcal{O}_{\mathcal{O}_{\mathrm{red}}}(X_{\mathrm{red}}) = k[x]$$

Prove that the complementary ideal is torsion-free and thus free, of rank 1 over $k[x]$. Let y be a generator.

Hint II-33. First, classify the primes by their codimension. Use factoriality to do the case of codimension 1. For codimension 2, show that a codimension 2 prime must contract to a codimension 1 prime in $\mathbf{Z}$, and then work modulo this contraction.

Chapter III

Hint III-11. One must, of course, work with open affine sets of the form $(\mathrm{Proj}\,S)_h$ with h of arbitrary strictly positive degree. There is still a one-to-one correspondence between the homogeneous primes of $S[f^{-1}]$ and the primes of $S[f^{-1}]_0$, but this requires a little attention.

Hint III-28. These double lines are all isomorphic projections of a double line in $\mathbf{P}_k^{d+2}$.

Hint III-32. Use Nakayama's lemma to go from a vector space basis of $(I(\mathcal{X})^v/I(\mathcal{X})^{v+1}) \otimes \kappa(p)$ to a set of generators for the A_p-module $(I(\mathcal{X})^v/I(\mathcal{X})^{v+1})_p$ and then to a set of generators for $I(\mathcal{X})^v/I(\mathcal{X})^{v+1}$ in a basic affine neighborhood of p.

Chapter IV

Hint IV-5. We have to show that for any $Z \subset X \times Y$ flat over Y, as in the exercise, the locus of points of Y such that the fiber of Z is nonreduced is closed; it is enough to take $Y = \mathrm{Spec}\,R$ affine. In this case, show that $Z = \mathrm{Spec}\,S$, where S is a quadratic extension of R, and show that the locus of nonreduced fibers is given by the discriminant.

In fact, the functors F and G of this problem are representable, by the *Hilbert scheme* $\mathcal{H}$ of degree 2 subschemes of X and by an open subscheme of $\mathcal{H}$, respectively (see IV-B-iv), so this could be considered a special case of Exercise IV-5; but we do not need to invoke this to do the exercise.

Hint IV-25. (Second part.) Look at the incidence correspondences

$$\Phi = \{(S, L) : S \text{ is smooth and } L \subset S\} \subset \mathbf{P}^{19} \times \mathbf{G}(1, 3)$$

and

$$\Psi = \{(S, L, C) : C, L \subset S\} \subset \Phi \times \mathcal{H}$$

Projection on the second factor shows that Ψ is irreducible of dimension 19; and projection on the first factor expresses Ψ as a projective bundle over Φ.

Hint IV-26. The residual intersection B will be a scheme of pure dimension 1 and degree 2; and we have a complete classification of these, given in III-B-iii. For each possible B (that lies on two or more quartics not having a common component and not both singular along $\mathrm{supp}(B)$), estimate the dimension of the family of pencils of quartics containing B.

References

Artin, M. (1971). *Algebraic spaces.* New Haven, CT: Yale University Press, Whitmore Lectures 3, 1969.

Atiyah, M. F., & MacDonald, I. G. (1969). *Introduction to commutative algebra.* Reading, MA: Addison-Wesley.

Bayer, D., & Eisenbud, D. (1991). *Ribbons.* (in preparation).

Bourbaki, N. (1989). *Commutative algebra.* New York: Springer-Verlag, Sections 1–7.

Brieskorn, E., & Knörrer, H. (1986). *Plane algebraic curves.* Boston: Birkhauser.

Collino, A. (1988). Evidence for a conjecture of Ellingsrud and Strømme on the Chow ring of $\text{Hilb}_d\mathbf{P}^2_{\mathbf{C}}$. *Illinois Journal of Mathematics, 32,* 171–210.

Cornell, G., & Silverman, J. H. (1986). *Arithmetic geometry.* New York: Springer-Verlag.

Danilov, V. I. (1978). The geometry of toric varieties. *Russian Mathematical Surveys, 33,* 97–154.

Deligne, P. (1974). La conjecture de Weil I. *Publications Mathématiques de l'Institute des Hautes Études Scientifiques, 43,* 273–307.

Demazure, M., & Gabriel, P. (1970). *Groupes algébriques: Vol. I. Géométrie algébriques, généralités, groupes commutatifs.* Paris: Masson.

Eisenbud, D. (1992). *Commutative algebra with a view toward algebraic geometry.* (in preparation).

Fong, L.-Y. (1991). *Rational ribbons and deformation of hyperelliptic curves.* (preprint).

Forster, O. (1981). *Riemann surfaces.* New York: Springer-Verlag.

Fulton, W. (1984). *Intersection theory.* New York: Springer-Verlag.

Godement, R. (1964). *Théorie des faisceaux.* Paris: Hermann.

Green, M. (1984). Koszul cohomology and the geometry of projective varieties (with an appendix by M. Green and R. Lazarsfeld). *Journal of Differential Geometry, 19,* 125–171.

Green, M., & Lazarsfeld, R. (1985). On the projective normality of complete linear series on an algebraic curve. *Inventiones Mathematicae, 83,* 73–90.

Grothendieck, A. (1957–1962). Fondements de géométrie algébrique. In *Séminaire Bourbaki.*

Grothendieck, A., & Dieudonné, J. (1960–1967). Eléménts de géometrie algébrique. *Publications Mathématiques de l'Institute des Hautes Études Scientifiques, 8, 11, 17, 20, 24, 28, 32.*

Gunning, R. C. (1990). *Introduction to holomorphic functions of several variables* (3 vols.): *Vol. III, Homological theory*. Pacific Grove, CA: Wadsworth and Brooks/Cole.

Harris, J. (1992). *Algebraic Geometry: A first course*. New York: Springer-Verlag.

Harris, J., & Eisenbud, D. (1982). *Curves in projective space*. Montréal: Les Presses de l'Université de Montréal.

Hartshorne, R. (1977). *Algebraic geometry*. New York: Springer-Verlag.

Hilbert, D. (1890). Über die theorie der algebraischen formen. *Mathematische Annalen, 36*, 473–534.

Hironaka, H. (1962). An example of a non-Kählerian deformation. *Annals of Mathematics, 75*, 190–208.

Iarrobino, A. (1985). Compressed algebras and components of the punctual Hilbert scheme. In *Algebraic Geometry, Sitges 1983*, E. Casas-Alvero (ed.). Springer Lecture notes in Mathematics 1124. New York: Springer-Verlag. 146–165.

Iitaka, S. (1982). *Algebraic geometry*. New York: Springer-Verlag.

Kempf, G., Knudsen, F., Mumford, D., & St.-Donat, B. (1973) *Toroidal embeddings I*. Springer Lecture Notes in Mathematics 339. New York: Springer-Verlag.

Knutson, D. (1971). *Algebraic spaces*. Springer Lecture Notes in Mathematics 203. New York: Springer-Verlag.

Kunz, E. (1985). *Introduction to Commutative Algebra and Algebraic Geometry*. Boston: Birkhauser.

Matsumura, H. (1986). *Commutative ring theory*. Cambridge, England: Cambridge University Press.

Milne, J. S. (1980). *Étale cohomology*. Princeton, NJ: Princeton University Press.

Mumford, D. (1962). Further pathologies in algebraic geometry. *American Journal of Mathematics, 84*, 642–648.

Mumford, D. (1963). Picard groups of moduli problems. In O. Schilling (Ed.) *Arithmetic algebraic geometry*. New York: Harper & Row.

Mumford, D. (1966). *Lectures on curves on an algebraic surface*. Annals of mathematical studies, Vol. 59. Princeton, NJ: Princeton University Press.

Mumford, D. (1967). Pathologies III. *American Journal of Mathematics, 89*, 94–104.

Mumford, D. (1976). *Algebraic geometry: Vol. I. Complex projective varieties*. New York: Springer-Verlag.

Mumford, D. (1988). *The Red Book of Varieties and Schemes*. Springer Lecture Notes in Mathematics 1358. New York: Springer-Verlag.

Mumford, D., & Fogarty, J. (1982). *Geometric invariant theory* (2nd ed.). New York: Springer-Verlag.

Nagata, M. (1962). *Local rings*. New York: Wiley Interscience.

Northcott, D. G. (1953). *Ideal theory*. Cambridge, England: Cambridge University Press.

Oda, T. (1985). *Convex bodies and algebraic geometry*. New York: Springer-Verlag.

Reid, M. (1990). *Undergraduate Algebraic Geometry*. Cambridge: Cambridge University Press.

Serre, J.-P. (1955). Faisceaux algébriques cohérents. *Annals of Mathematics, 61*, 197–278.

Serre, J.-P. (1979). *Local fields*. New York: Springer-Verlag.

Shafarevich, I. R. (1974). *Basic algebraic geometry*. New York: Springer-Verlag.

Silverman, J. H. (1986). *The Arithmetic of Elliptic Curves*. New York: Springer Verlag.

Spanier, E. (1966). *Algebraic topology*. New York: McGraw-Hill.

Swan, R. G. (1964). *The theory of sheaves.* Chicago: University of Chicago Press.

Thomas, A. D. (1977). *Zeta-functions: An introduction to algebraic geometry.* London: Pitman.

Vistoli, A. (1989). Intersection theory on algebraic stacks and on their moduli spaces. *Inventiones Mathematicae, 97,* 613–670.

Walker, R. J. (1978). *Algebraic curves.* New York: Springer-Verlag.

Zariski, O. (1947). The concept of a simple point on an abstract algebraic variety. *Transactions of the American Mathematical Society, 62,* 1–52.

Index of Symbols

$\mathcal{O}_X\|_U$	Sheaf $\mathcal{O}_X$ restricted to U,	23
$\mathfrak{m}_{X,x}$	Maximal ideal of $\mathcal{O}_{X,x}$,	24
$\mathrm{Hom}_{\mathscr{X}}(Y, Z)$	Set of morphisms from Y to Z in the category X,	121
ψ^*	Pullback by ψ,	26
$\mathbf{P}^n$	Projective n-space over k,	31
$X \times_S Y$	Fibered product of X and Y over S,	32
$\otimes$	Tensor product,	34
$\mathrm{Morph}_S(X, Y)$	Set of S-morphisms from X to Y,	37
$\mathbf{Q}$	Field of rational numbers,	38
$\mathbf{A}^n$	Affine n-space over k,	39
$\mathbf{Z}_p$	Ring of p-adic integers,	46
$\mathbf{Q}_p$	Field of p-adic numbers,	46
$\mathbf{C}_p$	Field of complex p-adic numbers,	46
$\mathbf{F}_q$	Field of q elements,	46
S_v	v^{th} graded piece of S,	88
$\mathrm{Proj}\, S$	Projective spectrum of S,	88
S^+	Irrelevant ideal of S,	89
$Z_+(I)$	Closed set of Proj S corresponding to I,	89
$H(X, v)$	Hilbert function of X,	103
$P(X, v)$	Hilbert polynomial of X,	103
$S(-b)$	Free graded S-module of rank 1, generated in degree b,	105
$\#X$	Underlying set of X,	120
(Hot)	Homotopy category of CW-complexes,	121
(schemes)°	Opposite category to the category (schemes) of schemes,	121
$T_p F$	Tangent space to F at p,	127
Λ^m	m^{th} exterior power functor,	132
$\mathscr{H}_p$	Hilbert scheme associated to a polynomial P,	133
$\mathscr{N}_{X/Y}$	Normal sheaf to X in Y,	137

Index